P9-BYF-028

The

TELEPHONE GAMBIT

ALSO BY SETH SHULMAN

Undermining Science

Unlocking the Sky

Owning the Future

The Threat at Home

The TELEPHONE GAMBIT

Chasing Alexander Graham Bell's Secret

Seth Shulman

W. W. NORTON & COMPANY
New York London

Printed in the United States of America
First Edition

For information about permission to reproduce selections from this book,
write to Permissions, W. W. Norton & Company, Inc.,
500 Fifth Avenue, New York, NY 10110

For information about special discounts for bulk purchases, please contact
W. W. Norton Special Sales at specialsales@wwnorton.com or 800-233-4830

Manufacturing by Courier Companies, Inc.
Book design by Chris Welch Design
Production manager: Andrew Marasia

Library of Congress Cataloging-in-Publication Data

Shulman, Seth.
The telephone gambit : chasing Alexander Graham Bell's secret / Seth Shulman. — 1st ed.
p. cm.
Includes bibliographical references and index.
ISBN 978-0-393-06206-9 (hardcover)
1. Bell, Alexander Graham, 1847–1922. 2. Gray, Elisha, 1835–1901.
3. Telephone—History. 4. Telephone—Patents. 5. Inventors—United States—Biography.
I. Title. II. Title: Alexander Graham Bell's secret.

TK6018.B4S58 2008
621.38509—dc22

2007030904

W. W. Norton & Company, Inc.
500 Fifth Avenue, New York, N.Y. 10110
www.wwnorton.com

W. W. Norton & Company Ltd.
Castle House, 75/76 Wells Street, London W1T 3QT

1 2 3 4 5 6 7 8 9 0

To my father,

with love and gratitude

CONTENTS

The TELEPHONE GAMBIT

PLAYING TELEPHONE

M*R.* W*ATSON—Come here*

Thomas Watson hunched over the bureau in Alexander Graham Bell's attic bedroom at the modest boardinghouse at 5 Exeter Place in Boston. Watson's ear pressed tightly against the metal frame of the small "speaking telegraph" receiver. His head faced the window. Outside, the city had grown dark and a full moon rose in the chill evening air.

The booming voice was unmistakable, even in a tinny, ghostlike facsimile. Watson reeled in amazement when he heard it. Jumping back, he swung open the bedroom door and ran into the hallway.

In the adjacent room, Bell was leaning over his workbench and shouting into the mouth of a metal cone clamped onto a block of wood. At the bottom of the cone, a piece of parchment was stretched tightly, like an upside-down drum. A platinum needle, stuck into a cork, was glued to the far side of the parchment from Bell's mouth. Its point dipped down into a small cup below that held a dilute solution of sulfuric acid.

When Bell yelled into the device, his bellowing voice vibrated the parchment diaphragm, slightly raising and lowering the needle into the solution and moving its tip alternately closer and further from a separate metal contact immersed in the cup. By attaching the top of the needle to a battery, Bell had created an electrical circuit that was completed only through the acidic water. The acid conducted electricity, but imperfectly. As a result, the vibration of the needle caused by the sound waves from Bell's voice correspondingly varied the resistance—or strength of the current—in the circuit. The machine looked something like this:

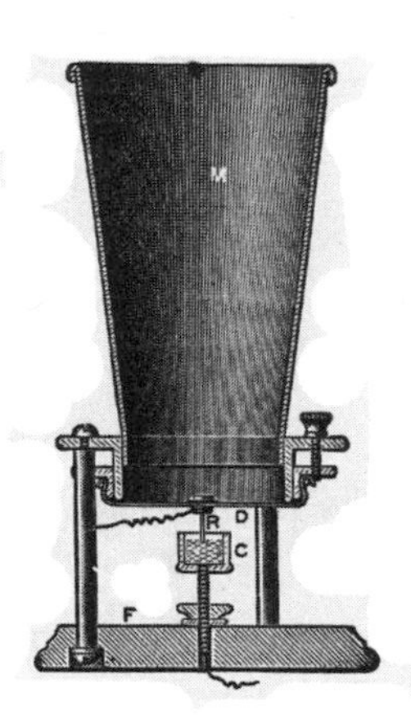

ALEXANDER GRAHAM BELL'S LIQUID TRANSMITTER. THE CUTAWAY VIEW (*RIGHT*) SHOWS THE SMALL CUP OF ACIDIC WATER BELOW THE METAL SPEAKING CONE.

With this pathbreaking liquid transmitter Bell had finally found a way to convert the sound waves of his voice into a signal that could be carried by electric current on a wire. It was a brilliant and elegant invention—and a radical departure from his previous research.

Wires from this novel, battery-driven liquid transmitter snaked their way down the hall to the bedroom next door, where a telegraphlike receiver was fitted with a small, vibrating strip of metal. Bell had used the receiver part of the apparatus many times before in a series of experiments over the past year and a half. In his terminology, the sensitive metal "reed" was intended to vibrate in concert with the "undulations" created in the electrical current.

Bell was undoubtedly still shouting into the contraption when Watson burst into the room to report what he had heard. Only then did Bell realize that he had placed the world's first telephone call.

The feat was hard for either man to believe. They had labored toward the goal for so long—and with such paltry success—prior to that moment on March 10, 1876.

THE BELL AND WATSON "eureka moment" is one of the best-known stories in the history of invention, and among the most romantic, with its two earnest and visionary young inventors profoundly changing the world from their humble quarters. I first heard the tale as a child and have read enough versions now for its familiarity to give it a luster, the way polished old wood develops a patina.

Bell's voice traveled just ten yards along a bare wire from one room to another. It was a modest transmission of sound waves to have unleashed such an enormous change in human interaction. Yet even today, after untold billions of long-distance conversations spanning more than a century, Bell's oddly emphatic words are still probably the most famous ever uttered into a telephone.

The story of the telephone's invention is not just well known, it is impeccably documented, beginning that very evening, when both men wrote about it in their notebooks. With contemporaneous eyewitness accounts from each of the episode's principals, the story is the historian's equivalent of a slam dunk.

Bell's formal account is written in neat script in the pages of his leatherbound laboratory notebook, beneath his detailed diagram of the liquid transmitter:

> *I then shouted into* **M** *[the mouthpiece] the following sentence: "Mr. Watson—Come here—I want to see you." To my delight he came and declared that he had heard and understood what I said. I asked him to repeat the words. He answered "You said—'Mr. Watson—come here—I want to see you.' " We then changed places and I listened at* **S** *[the reed receiver] while Mr. Watson read a few passages from a book into the mouth piece* **M**. *It was certainly the case that articulate sounds proceeded from* **S**. *The effect was loud but indistinct and muffled. . . . I made out "to" and "out" and "further" and finally the sentence "Mr. Bell do you understand what I say? Do–you—un—der—stand—what—I—say" came quite clearly and intelligibly.*

Watson's notes are more cryptic and hastily penned, in now-faded grayish ink toward the beginning of a tall, slim notebook of which only the first twenty pages are filled. Noting the date, Watson jotted down some of the phrases he and Bell tried to pass along the wire to one another. *Mr. Watson come here I want you,* he wrote on one line. Below it:

> *How do you do.*

The next line reads:

> *God save the Queen and several other articulated sentences.*

Unlike some other inventors at their moment of discovery, Bell and Watson recognized at once the implications of the momentous threshold they had crossed. Bell was elated. In a letter to his father that same night, he waxed prophetic, writing:

> *I feel that I have at last found the solution of a great problem and the day is coming when telegraph wires will be laid on to houses*

> *just like water or gas is, and friends will converse with each other without leaving homes.*

Much later, Watson even quietly pocketed the wires that had carried the pioneering conversation. He reverently coiled them up almost as though they were a religious relic. Watson placed the wires in an envelope and brought them to a safe-deposit box at the bank. He attached a note to the coiled wires that read:

> *This wire connected Room No. 13 with Room No. 15, at 5 Exeter Place, and is the wire that was used in all the experiments by which the telephone was developed, from the fall of 1875 to the summer of 1877, at which latter time the telephone had been perfected for practical use. Taken down July 8th, 1877,*
>
> *Signed, T. A. Watson.*

Bell's and Watson's notebooks and even the wires they used have survived to this day. But history, I have come to learn, is an odd business. We can know a tremendous amount about what happened in the past and yet still understand very little.

SEVERAL YEARS AGO, in 2004, I accidentally stumbled upon a trail of information that casts a shadow over the now-famous and familiar story of Bell and Watson. It started as a nagging question that occurred to me when I read Bell's laboratory notebook. But it built into something stranger. Over the course of my research, Bell and Watson's tale, told for generations and written into textbooks, transformed from a beloved and earnest "eureka moment" to something more complex and disturbing.

There is little doubt that Bell's call to Watson is now the best-known story about him. I would become convinced—much to my own amazement—that the simple tale also immortalizes what was likely the most ignominious act of Bell's life. In some sort of cosmic historical irony, the well-known tale that makes Bell's name also freezes him for all time in the midst of one of the most consequential thefts in history.

As I investigated the roots of Bell's story of invention, much of what I thought I knew about him came into question. I would ponder, at some length, how so many historians could have gotten the facts of Bell's story so wrong. And I would wonder, especially at times of frustration, whether history isn't a bit like the children's parlor game of "telephone" where one person whispers into the next person's ear until the starting phrase becomes twisted beyond recognition.

But I am getting ahead of my story. Let me just say this: stranger things have happened—but not to me. Absolutely by accident I fell through a kind of historical trapdoor into a vexing intrigue at the heart of one of the world's most important inventions.

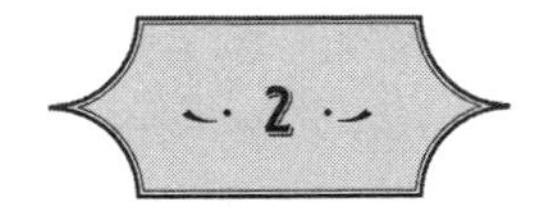

DISCONNECTED

LATE ONE OCTOBER evening, I was working in the plush office I had been given for the year at MIT. On my computer screen, courtesy of the Library of Congress, was a high-resolution, digital reproduction of Alexander Graham Bell's laboratory notebook from 1875–76, exactly as he had written it in his own hand.

The large windows near my desk looked out on the Charles River and downtown Boston. I gazed at the glow of the night skyline and the headlights of cars speeding into and out of the city. I realized I could practically see the spot at 5 Exeter Place where, more than a century ago, Bell had written the words before me.

On the screen, the images of Bell's notebook lacked only the musty smell of its leather binding and the brittle feel of its lined pages. In every other respect, they offered a perfect facsimile, allowing the viewer to follow Bell's work straight from his own fountain pen. In some passages, I thought I could even roughly gauge Bell's excitement from the way his script got scratchy when he wrote more hurriedly.

I wondered what Bell would have made of the fact that I was viewing a perfect reproduction of his notebook via the World Wide Web. He'd surely marvel at the technology. And he would also be justified to feel proud. After all, the Internet is little more than a powerful descendant of the communication device he himself pioneered.

As a journalist who specializes in science and technology, I have long been interested in invention—how it occurs and how it is remembered. So I jumped at the opportunity to spend a year as a science writer-in-residence at the Dibner Institute for the History of Science and Technology at MIT. It was the first time they had invited an outsider to join in the program's seminars and discussion groups. And it was my first experience working alongside a group of historians.

Given my interest in inventors, I had proposed to do a year of research on the relationship between two towering icons: Thomas Alva Edison and Alexander Graham Bell. It was a project that seemed full of possibility and I was grateful for the opportunity to begin it. Edison and Bell are, of course, renowned for the world-shaping technological contributions they made early in their careers. By age thirty-four, Edison had been dubbed the "wizard of Menlo Park" for his work on the incandescent lightbulb. Bell became famous for the telephone by the age of thirty.

Less commonly known is the fact that Edison and Bell were close contemporaries, born just twenty days apart in 1847. Or that they were long-standing, often bitter rivals, with dramatically divergent temperaments, backgrounds, and approaches to invention. Edison, a cantankerous autodidact, had just three months of formal schooling. He worked slavishly long hours at his factorylike lab, made his assistants punch a time clock, and stressed experimentation and persistence above all else. Edison is always popularly associated with the incandescent lightbulb, but he is also responsible for a host of other pathbreaking inventions. His 1,093 patents set a record that still stands: the most ever garnered by a single individual. Every bit as impressive as the sheer scale of his productivity is its breadth. Edison's inventions include motion pictures, the phonograph, the electric car, even the use of poured concrete in building construction, to name just a few.

Edison doubtless accomplished more, but Bell wins points for panache. Having studied at University College in London and taught at Boston University, he was a refined, aristocratic professor, who slept late, liked to play the piano in the evenings, and believed that theory should guide experimental research. Bell projected an enlightened sensibility, and his interests, scientific and otherwise, ranged even further than Edison's did. Bell not only compulsively followed technological developments in several languages, he liked to end his day by reading entries from the encyclopedia. He opened up progressive new avenues for teaching the deaf and introduced Helen Keller to her life-changing teacher. He helped launch the journal *Science* in the 1880s and served as president of the National Geographic Society. After the telephone, Bell even worked with a small team to design one of the world's first successful airplanes. Bell's concern over congenital deafness led to a sorry foray into the eugenics movement of his day, but overall he was broad-minded in his support for causes such as women's rights and universal suffrage.

Equally notably, most everyone seemed to like Bell—for his warmth, his erudition, and especially for his seemingly indefatigable verve for all things new. Perhaps one of the best accounts to capture Bell's beloved manner comes from his son-in-law, David Fairchild. As he recalls,

> *Mr. Bell was tall and handsome with an indefinable sense of largeness about him, and he so radiated vigor and kindliness that any pettiness of thought seemed to fade away beneath his keen gaze. He always made you feel that there was so much of interest in the universe, so many fascinating things to observe and to think about, that it was a criminal waste of time to indulge in gossip or trivial discussion.*

I was fascinated by Edison's accomplishments. But, given what little I knew, I felt more of an affinity for Bell, so I decided to begin my research with him. I started to fill my office shelves with a large collection of secondary sources from the institute's library about Bell's life and times. But his own beautifully detailed laboratory notebooks offered my most natural point of departure.

I COULDN'T HAVE come to a better place for this kind of work. The Dibner Institute offered a rarefied haven for scholars from around the world. Tucked into the westernmost corner of the bustling mosaic of the MIT campus, the institute was bequeathed by Bern Dibner, an enigmatic industrialist who, in the warehouse of his factory in Bronx, New York, amassed one of the world's foremost collections of books on the history of science and technology. As Dibner's interests turned increasingly from his business (making electric connectors such as those used for high-voltage power lines) to his passion (the history of technology), his warehouse shelves burgeoned with rare volumes by everyone from Archimedes to Volta.

Each year since its founding in 1990, the Dibner Institute had invited a few dozen scholars to set up shop in its luxurious and decidedly unwarehouselike headquarters, to use its famed library, and to conduct research. In the office next to mine, Cesare Maffioli, a genial Italian, was studying Leonardo da Vinci's uncanny grasp of hydrodynamics. Peter Bokulich, a younger colleague on the floor below with expertise in both physics and history, was investigating how one particular scientific article—known as the Bohr-Rosenfeld paper—influenced the emergence of the field of quantum mechanics. Not my normal cohort, to be sure, but a very interesting group to get to know.

My first surprise when I opened the modern glass doors to the institute's front office was the large oil portrait of Alexander Graham Bell hanging prominently over the receptionist's desk. In the picture, Bell stands beside a workbench or table, white-haired and beneficent-looking, gazing thoughtfully into the distance. The front office also had elegant, modern glass displays lining the walls that held choice artifacts from the history of technology. One display case held a negative from an early Roentgen X-ray machine. Another featured an early Morse telegraph in mint condition. Sitting near it was a handsomely bound period copy of Bell's first public speech about the telephone. The caliber of the well-chosen artifacts telegraphed its own message: lest there was any doubt, this place took the subject of invention seriously.

It was comforting to see Bell, the subject of my research, so visibly affirmed in the institute's front office, because almost all of my colleagues for the year were academic historians and I was aware from the first of my status as an outsider to the ivory tower.

Not long after my arrival, George Smith, the acting director of the program, knocked on my office door. Professor Smith, a warmhearted, erudite scholar with big glasses and a shy smile, had helped design jet engines as a young engineer. Qualms of conscience about his role in military research led him to the field of philosophy. After many years of scholarly work and teaching, he had acquired an encyclopedic knowledge of the history of science and technology and emerged as one of the world's leading experts on the life and work of Sir Isaac Newton.

Smith welcomed me and he even graciously cited several of my writings. He said he was particularly pleased by a book I had written on the early history of aviation because he felt it rightly emphasized the work of a pioneer too often neglected. Plus, he said, he was happy to have someone at the institute who wrote for an audience beyond the narrow world of academic historians.

Smith was quick to note, however, that not everyone shared his views. In an oddly friendly warning, he told me that I would have to prove myself to some of the historians at the institute.

"Quite frankly," he said, "you should know that there are a lot of people who don't think you should be here."

ON THE NIGHT I was paging through Bell's notebook, though, I wasn't worried about what anyone else thought. I was simply engrossed in Bell's research. The first thing I noticed was its sensible progression. Day after day, Bell made incremental changes in his experiments using the same elements: electromagnets, vibrating reeds, and tuning forks. The work was clear, tangible, elegant. And compared to a lot of the modern-day science I have covered as a reporter, I felt I understood what Bell was doing and maybe even what he was thinking about.

For the most part, he was not thinking about the device we now call the telephone, which of course did not yet exist. Rather, Bell, like many other inventors of his day, was primarily trying to solve a problem then plaguing the burgeoning telegraph industry: how to send more than one message at a time over a wire. Bell didn't know too much about the relatively new fields of electricity and magnetism, but he was a well-schooled teacher of the deaf and, as such, he did know a good deal about sound. Bell's idea for what he called a "multiple telegraph" was to try to send and receive messages at different musical pitches, or frequencies. If he could build devices tuned to send and receive messages at one single musical pitch, he reasoned, he might be able to send multiple messages over the same telegraph wires simultaneously without having them interfere with one another.

Toward that end, Bell was systematically trying to build a series of telegraphlike devices that would be receptive only to a signal sent at a particular pitch, or frequency. He tried batteries of different strengths. He tried magnets in different arrangements. He even built a cylinder lined with bar magnets that could be spun at different speeds to adjust the pitch of the vibrating reeds in his circuit.

Of course, in addition to this commercial goal, Bell was interested in many other things, too. He had a sharp, restless mind and a great imagination. He was fascinated by the possibility of transmitting vocal sounds over telegraph wires. And he was so interested in the way people perceived sound that, with the help of Clarence Blake, a local doctor, he even experimented on the ear of a human corpse as part of his work during this period. Bell's grasp of acoustics and his focus on try-

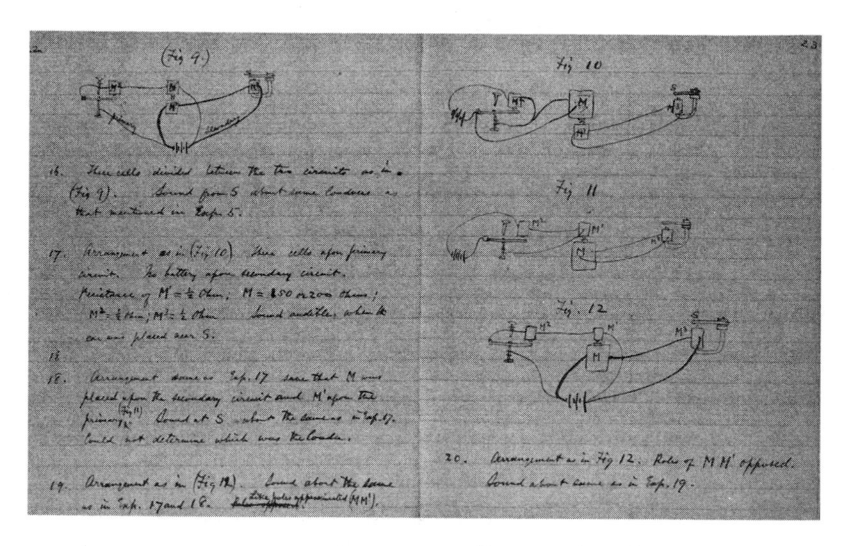

(Fig 9)

16. Three cells divided between the two circuits as in (Fig 9). Sound from S about same loudness as that mentioned in Exp. 5.

17. Arrangement as in (Fig 10). Three cells upon primary circuit. No battery upon secondary circuit. Resistance of M' = ½ Ohm; M = 150 or 200 Ohms; M² = ½ Ohm; M³ = ½ Ohm. Sound audible, when the ear was placed near S.

18. Arrangement same as Exp. 17 save that M was placed upon the secondary circuit and M' upon the primary (Fig 11). Sound at S about the same as in Exp. 17. Could not determine which was the louder.

19. Arrangement as in (Fig 12). Sound about the same as in Exp. 17 and 18. ~~[illegible]~~ Like poles approximated (MM').

Fig 10

Fig 11

Fig. 12

20. Arrangement as in Fig 12. Poles of M M' opposed. Sound about same as in Exp. 19.

A SAMPLE PAGE FROM ALEXANDER GRAHAM BELL'S NOTEBOOK FROM 1875–76. THIS SPREAD (PP. 22–23), WHICH IS DATED FEBRUARY 22, 1876, SHOWS SOME IDEAS FOR A MULTIPLE TELEGRAPH DESIGN.

ing to send differently pitched sounds over the telegraph wires would quickly lead him to envision a device that could successfully transmit the human voice.

I lost track of how late it was getting. Somewhere around midnight, I reached Bell's accounts from March 1876, the period of his momentous breakthrough with the telephone. One entry jumped out at me.

Bell's research notes on March 8 shift to some strikingly new ideas after months of slow, incremental work. On that day, for the first time, Bell inexplicably adds to his experiments a dish of water laced with sulfuric acid. He still uses a reed and a magnet at one end of the circuit he is building, but, seemingly from nowhere, he introduces a striking new contraption: a diaphragm with a needle sticking through it into the acidic water to complete the electrical circuit. From that entry on, some liquid or another becomes a feature in a quick succession of experiments. And, of course, just a day and a half after introducing this new scheme, Bell has his amazing success calling to Watson next door.

After following many months' worth of Bell's steady and methodical work, I couldn't help but be struck by his sudden conceptual leap. He had recorded nothing like it over the course of the preceding year. What made Bell think of dipping a needle into liquid in his transmitter, I wondered, after a steady diet for more than a year of reeds, magnets, and batteries in widely varied configurations?

I viewed the shift as a sign of Bell's genius. I made a note to that effect in my own handwritten journal that night.

I was especially struck by this shift in Bell's thinking because I've found that a kind of magic often seems to inhabit the moment of discovery: that instant when something formerly unknowable, beyond reach, becomes forever clear. The element that makes such a shift possible—the fleeting insight or fortuitous accident—is often hard, if not impossible, to explain or capture. I'm as excited by such moments as a prospector might be to unearth a rich vein, or a book collector to stumble upon a vanishingly rare first edition.

Wilbur Wright's pathbreaking idea for bending the wings of an airplane to give it control in the air purportedly came to him in his bicycle shop while he idly twisted a box that had held an inner tube. Alexander Fleming, excellent scientist that he was, became fascinated by the mold that had crept in from the damp London air to ruin his experiment growing colonies of *Staphylococcus* bacteria in culture. Thankfully, Fleming studied the mold instead of tossing the tainted samples, and penicillin was the world-changing result.

Musing on the account in Bell's notebook of his experience at the threshold of his discovery, I had a little "eureka moment" of my own. I noticed that there was a twelve-day gap between Bell's entries at the end of February and those beginning in March. With consecutive entries on every page, notebooks often make work appear relatively seamless; but in this case, Bell left his experiments on February 24 and didn't resume them for nearly two weeks. He mentions the fact clearly himself on page 34, the day before he introduces his new transmitter idea, with this succinct notation:

Returned from Washington March 7th, 1876.

It seemed immediately clear that Bell's absence from his lab spurred him in a new direction. I wanted to find out more about the trip to Washington and what might have led him to change his work so noticeably upon his return.

This, I imagined, was precisely the kind of challenge my historian colleagues engaged in routinely. For me, though, it was new. A primary source document had suggested a small, historical puzzle. And I was fortunate enough to have the time and resources available to explore such an open-ended lead. I doubted the question had much to do with Bell's rivalry with Thomas Edison. But I was curious, nonetheless. I never for a moment suspected what would happen next.

ON THE HOOK

On Saturday morning, February 26, 1876, Alexander Graham Bell's train pulled into the Baltimore & Potomac Railroad station in downtown Washington, D.C., with a loud hiss and metallic squeal. Bell was agitated as he hurriedly stepped down onto the station platform. He was keenly aware that his fortune hinged on the outcome of the events before him—almost all of which were beyond his control.

Aleck Bell, as he was then known, was days away from his twenty-ninth birthday, an intensely driven and serious young man with wavy black hair and a scruffy beard. A teacher of the deaf and an associate professor at fledgling Boston University, Bell was a man of meager means but grand ambitions. Acquaintances in this period remember Aleck Bell's studied manners and speech. They recall, too, that his formal demeanor—from his straight-backed posture to his schooled diction—made him seem much older than he actually was.

Leather suitcase in hand, Bell made his way to the station entrance. It was a pivotal moment for Bell and an exciting time for a nation nearing

Alexander Graham Bell in 1876.

its centennial year, with an unsettling, frontierlike quality to many of the changes under way. As Bell knew well, nothing illustrated the point better than the railroads themselves. In the past four years alone the United States had laid an astonishing 12,000 miles of track, speeding the nation headlong into a new era. And everyone, just like Bell himself, seemed to be disembarking into a bewildering and exciting new world of mechanical and electrical devices, from sewing machines to fire alarms; of grand new scientific ideas like Charles Darwin's evolutionary theory. A world beset by hucksters like P. T. Barnum with his "Greatest Show on Earth"; by crooks like William "Boss" Tweed, who had plundered New York City as its elected official and recently escaped jail to flee the country; and even outlaws like Jesse James and his gang, who had not long ago held up passengers in an audacious raid on the Rock Island Line.

Outside the station, the weather was unseasonably warm. A fine carriage and driver met Bell to deliver him to the home of his wealthy and powerful patent attorney, Anthony Pollok. Bell would be Pollok's guest during his visit.

THE U.S. PATENT OFFICE IN WASHINGTON, D.C., CIRCA 1876.

Out the window of Bell's carriage, the nation's capital had an unfinished and rough-edged air, especially compared with his adopted hometown of Boston. The broad avenues were sparsely settled; cheap and shabby hotels and shops stood near grand government buildings. Many roads had yet to be paved. When Charles Dickens had toured Washington a few decades earlier, he declared it little more than a pretentious village, calling it the "city of magnificent intentions."

The capital had doubled in population since then, but now, at the tail end of President Ulysses S. Grant's second term, Dickens's words still echoed. As if to reinforce the sentiment, the half-finished Washington Monument rose from the center of town like an unsightly exclamation point, its construction stalled for nearly three decades since before the Civil War. President James Polk had famously laid the cornerstone in 1848; but now, in 1876, the monument still stood like an oversized broken-off chimney, replete with a makeshift roof pitched over the top to keep the rain out.

Unlike the provisional feel of many parts of the city, however, Bell found Pollok's home to be opulent. As Bell wrote his father during his stay:

> *Mr. Pollok has the most palatial residence of any that I have ever seen. It is certainly the finest and best appointed of any in Washington.*

Among its amenities, Pollok's Gilded Age mansion boasted granite pillars, fifteen-foot-high ceilings, and, Bell noted, a large staff of "colored servants." During Bell's stay, Pollok would introduce him to many people in Washington's high society and even host a party in his honor.

Bell was too preoccupied with his own affairs, though, to dwell much upon any of this. As he confided to his father,

> *You can hardly understand the state of uncertainty and suspense in which I am now.*

As Bell put it, his entire future rested on the outcome of the "patent muddle" he had come to Washington to sort out. The stakes were high. Bell's telephone patent—a claim that would come to be known as the most lucrative patent in history—had been threatened with a formal declaration of "interference," the term the U.S. Patent Office uses when two or more inventors apply for patents on overlapping inventions at the same time. Now, the patent examiner thought he might have found more overlap between Bell's claims and those of others. The finding raised the prospect of potentially protracted and expensive interference proceedings.

Bell knew that such a dispute was to be avoided at all cost. It could drag on for years, during which it would stall his ability to reap any financial reward from his work for himself and his financial backers. Only a clear and unfettered patent would allow Bell, a relatively unknown outsider to the telegraph industry, to successfully commercialize his research.

Bell laid out the situation clearly in a letter to his father from Washington. He saw it as a key turning point in his life. Bell wrote that if he lost his patent bid, he would abandon his electrical research altogether and devote himself full time once again to teaching the deaf. But if he succeeded in winning his patent claim, he would feel confident enough to marry the wealthy young woman to whom he had recently become engaged. As he put it excitedly, with the emphasis as it appears in his letter:

> *If I succeed in securing that patent without interference from the others,* the whole thing is mine . . . *and I am sure of fame, fortune, and success if I can only persevere in perfecting my apparatus.*

OVER THE NEXT few days, in a binge of my own work, I puzzled over a complex knot of irregularities about Bell's life-altering visit to Washington, D.C. To begin with, the timing of the trip seemed more than a little odd. Bell filed his telephone patent on February 14, 1876, but, according to his laboratory notebook, he did not successfully transmit intelligible speech over a telephone until March 10. Was it true that, in the lingo of the Patent Office, Bell had yet to "reduce his invention to practice" at the time he filed his patent application? That, in other words, Bell patented an invention he had never actually made?

Even the logistics of this question were mystifying. I knew from reporting on disputes over intellectual property that working models of inventions were required by the U.S. Patent Office in the 1800s. It took only a little digging to learn that on February 14, 1876—the very day Bell filed his telephone patent—a U.S. Senate committee held hearings on a bill calling for the agency to do away with this requirement. Supporters of the bill, proposed by Connecticut senator James E. English, testified that the Patent Office's attic coffers were literally overflowing and that there was no space to put the roughly twenty thousand new models the agency expected to receive in the coming year.

Of course, there would have been no point for the Senate to debate the issue unless, by February 1876, the Patent Office at least technically continued to require working models to accompany patent applications. Why, then, hadn't the patent examiner in Bell's case required him to submit a functioning model of his telephone?

Equally baffling was the Patent Office's decision to swiftly grant Bell his telephone patent before he had even returned to his lab in Boston on March 7, 1876. How was it, I wondered, that one of the most momentous patents in history was issued in just three weeks? When I looked up other patents filed or issued around the same time, they all seemed to have taken months, if not years, to issue. Timothy Stebins, another little-known Boston-based inventor, had filed a patent for a hydraulic elevator on March 2, 1876, but it wasn't granted until more than five months later, on August 15. William Gates, a New Haven–based inventor with a newfangled electric fire alarm, found that his patent application took almost two years to process. It was filed on April 1, 1874, but the patent wasn't issued until the time of Bell's visit on February 29, 1876. A little more research offered a likely explanation: in 1876, the U.S. Patent Office employed just a few dozen patent examiners to process the *tens of thousands* of applications the agency received each year.

The U.S. Patent Office's speedy work to approve Bell's patent seemed all the more extraordinary because, on February 19, 1876, the patent examiner had notified Bell that his patent would be "suspended" for three months, after which time, the letter said, the office would formally decide whether to declare so-called interference proceedings. Such interference disputes almost always include formal hearings to determine which inventor can rightfully claim "priority of conception." Sorting out the interference claims on inventor Emile Berliner's 1877 patent application on the microphone, for instance, ended up taking more than thirteen years. That was, of course, an extreme case, but even the more common interference proceedings were likely to last for one or more years.

There was no question about it: the swiftness of the U.S. Patent Office's actions seemed highly unusual. I wondered what had made patent officials change their minds so quickly about their contention that the claims of others overlapped with Bell's. For that matter, I wondered exactly what those other claims were.

Thanks to the Dibner Institute's extraordinary library at MIT, I could easily answer the second question. The filing that conflicted with Bell's telephone patent came from an electrical researcher named Elisha Gray.

TODAY, IF HE is remembered at all, Elisha Gray is known as a technological footnote: the unlucky sap whose patent claim for a telephone arrived just hours after that of Alexander Graham Bell.

History is harsh in ascribing winners and losers.

As I soon learned once I started looking, there is a good deal of information to be had about the fight between Bell and Gray over rights to the telephone. The battle dragged on through the courts, in one form or another, for more than a decade. But it is not much remembered today. After all, there is little question about who prevailed in the end.

Ironically, though, back in 1876, Gray was far better known than Bell. Some twelve years Bell's senior, Gray was recognized, at least in scientific circles, as one of the leading electrical researchers in the country. He had already received enough money and acclaim for his

Elisha Gray.

work to devote himself full time to inventing, and had received many of the roughly seventy patents around the world he would ultimately garner for his inventions—far more than Bell would ever claim.

Gray was born in 1835 on an Ohio farm; when he was twelve, his father died, plunging the family into poverty. Gray had to quit school to go to work. Despite his lack of formal education, though, he became fascinated by the seemingly magical new possibilities promised by electricity in the mid-1800s. Coupling his fascination with resolve, Gray managed to support himself as a carpenter while completing preparatory school and two further years of study at Oberlin College, near his home. Then, in 1868, at age thirty-three, Gray received his first patent—for an improved telegraph relay. Initially, Gray conducted his electrical experiments in addition to farming. Building upon his patent's success, however, he soon helped start a firm called Barton &

Gray to manufacture telegraphic equipment and he launched a full-time career as a manufacturer and inventor.

Before long, Western Union, the vast U.S. company that held a near monopoly over the telegraph, recognized the impressive caliber of Gray's work. In 1872, the company bought a one-third interest in Gray's firm, making him a wealthy man. Gray's company changed its name to Western Electric, moved to Chicago, and soon became the leading developer and supplier of equipment to Western Union.

By 1876, the quality of Western Electric's products was universally admired in the emerging field. For instance, Bell's assistant Thomas Watson vividly recalls how he and the others in the Charles Williams machine shop (where Watson crafted Bell's telegraph inventions) looked upon Gray's work with at least a tinge of envy:

> *Gray was electrician for the Western Electric Company of Chicago, the largest manufacturer of electrical machinery in the country at that time. His shop had better tools and did finer work than Williams'. Whenever a piece of Western Electric machinery came into our shop for repairs, the beauty of its design and the quality of its workmanship made it an object of admiration to all of us, and made most of Williams' instruments look crude.*

In the case of the telephone, I learned, Gray had filed what the Patent Office then called a "caveat." Although the government subsequently dropped the option by 1910, a caveat issued by the U.S. Patent Office provided an inventor up to a year with an exclusive right to turn his or her idea into a working, patentable invention. In those days an inventor who had conceived of a device but had yet to build it could use a caveat to warn away would-be competitors. Once it was granted, a caveat afforded all the same rights as a patent during the provisional year while the applicant worked to complete the invention in question. Gray's caveat described an "instrument for transmitting and receiving vocal sounds telegraphically." In what is normally described as a strange twist of fate, Gray filed his claim on February 14, 1876—the very same day Bell filed his patent application.

As I inspected the caveat document, reprinted in a book on the history of the telephone, I learned that Gray proposed to use a liquid in his telephone transmitter: water with acid in it. That fact alone seemed like a remarkable coincidence.

But Gray's sketch for his invention, on page 3 of his patent claim, hit me almost like a shock from the electric current it described. I recognized immediately that I had just seen a virtually identical drawing—in Bell's lab notebook.

The implication was instantly clear. Unless I was somehow mistaken, Bell must have returned to his lab in Boston from his trip to Washington, dropped his prior line of inquiry, and drawn an almost perfect replica of his competitor's invention in his own notebook.

As I stared incredulously at the drawing in Gray's caveat, I tried to make sense of the chain of events. Gray had filed a confidential caveat at the U.S. Patent Office, clearly outlining his prescient idea for a machine to transmit speech, an invention he had envisioned fully but had yet to build. Bell, on the other hand, returned from a visit to the nation's capital in possession of a U.S. patent on an invention that had never yet transmitted speech. Upon his return to Boston, Bell scrapped his former efforts and sketched an unmistakable picture of his competitor's idea for a liquid transmitter in his own laboratory notebook, passing it off as his own discovery. Next, in his laboratory in a boardinghouse on Exeter Street, Bell built and used this machine—Gray's machine—to carry on what would forever be immortalized as the world's first telephone conversation.

I was dumbfounded. Could Bell have committed such a blatant, wholesale act of plagiarism? If he did, I wondered, how could no one have noticed it before? After all, however long ago it may have occurred, this was an act of tremendous historical consequence. The telephone sits high atop any list of the most important modern inventions, and Alexander Graham Bell is surely one of the best-known inventors of all time. Even beyond issues of fame and historical accuracy, Bell's seemingly iron-clad patent claim to the telephone led directly to a company, American Telephone & Telegraph, that would become one of the largest and most lucrative monopolies the world has ever known.

I know it sounds improbable that Alexander Graham Bell, almost universally canonized as the inventor of the telephone, might be undeserving of the title. Or that I, in a relatively casual reading of Bell's notebook, might have discovered something that had eluded generations of historians. So, before going on with my tale, let me pause a moment for those who, reasonably enough, suspect that my account is fictionalized or embroidered. Here, for your own inspection, are the documents that first set me upon the strange quest to track down the true story about Alexander Graham Bell:

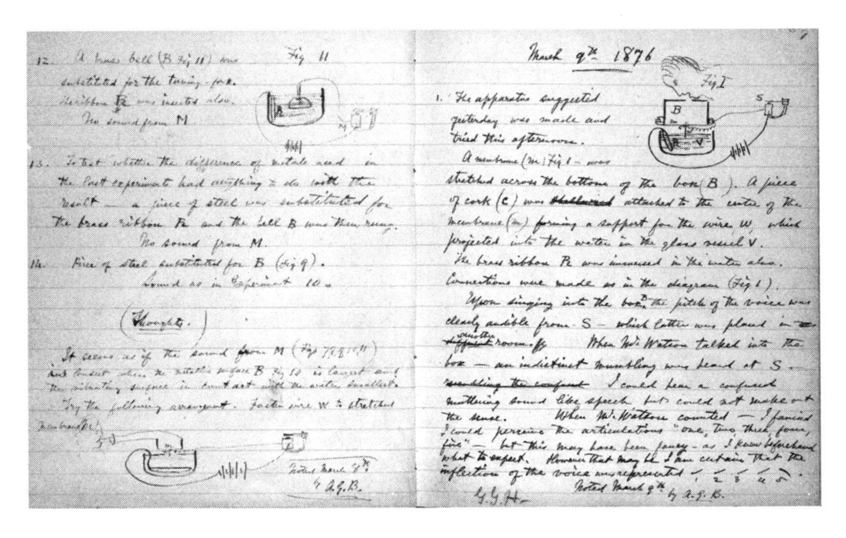

12. A brass bell (B Fig 11) was substituted for the tuning-fork.
The ribbon R was inserted also.
No sound from M.

Fig 11

13. To test whether the difference of metals used in the last experiment had anything to do with the result — a piece of steel was substituted for the brass ribbon R and the bell B was then rung.
No sound from M.

14. Piece of steel substituted for B (Fig 9).
Sound as in Experiment 10.

(Thoughts.)

It seems as if the sound from M (Fig 7, 8, 10, 11) was loudest when the metallic surface B Fig 10 is largest and the vibrating surface in contact with the water smallest.
Try the following arrangement. Fasten wire W to stretched membrane (M)

Noted March 8th by A.G.B.

March 9th 1876

Fig 1

1. The apparatus suggested yesterday was made and tried this afternoon.
A membrane (M; Fig 1 — was stretched across the bottom of the box (B). A piece of cork (C) was attached to the centre of the membrane (M) forming a support for the wire W, which projected into the water in the glass vessel V.
The brass ribbon R was immersed in the water also.
Connections were made as in the diagram (Fig 1).
Upon singing into the box B the pitch of the voice was clearly audible from S — which latter was placed in another room. When Mr Watson talked into the box — an indistinct mumbling was heard at S. I could hear a confused muttering sound like speech but could not make out the sense. When Mr Watson counted — I fancied I could perceive the articulations "one, two, three, four, five" — but this may have been fancy — as I knew beforehand what to expect. However that may be I am certain that the inflection of the voice was represented 1 2 3 4 5.
Noted March 9th by A.G.B.

THE PAGE FROM ALEXANDER GRAHAM BELL'S NOTEBOOK DATED MARCH 9, 1876. BELL HAD JUST RETURNED FROM VISITING THE U.S. PATENT OFFICE IN WASHINGTON, D.C., AND WOULD THE VERY NEXT DAY RECORD HIS SUCCESS WITH THE WORLD'S FIRST TELEPHONE CONVERSATION.

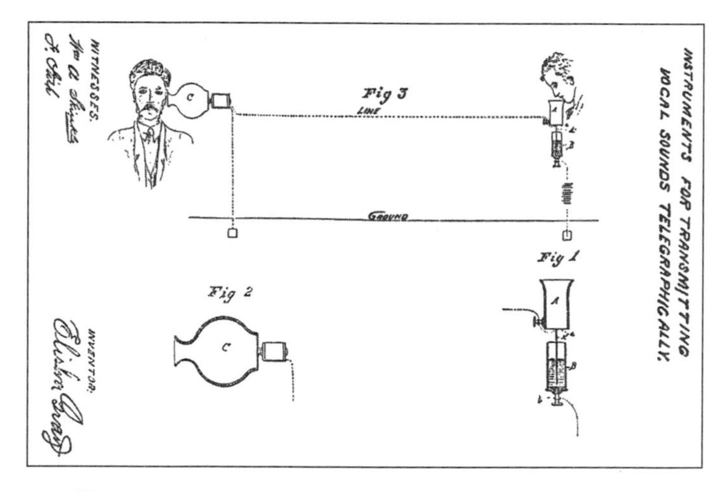

THE DIAGRAM FROM PAGE 3 OF ELISHA GRAY'S CAVEAT, A CONFIDENTIAL DOCUMENT FILED AT THE U.S. PATENT OFFICE ON FEBRUARY 14, 1876, ALMOST THREE WEEKS BEFORE A NEARLY IDENTICAL DRAWING APPEARED IN BELL'S NOTEBOOK.

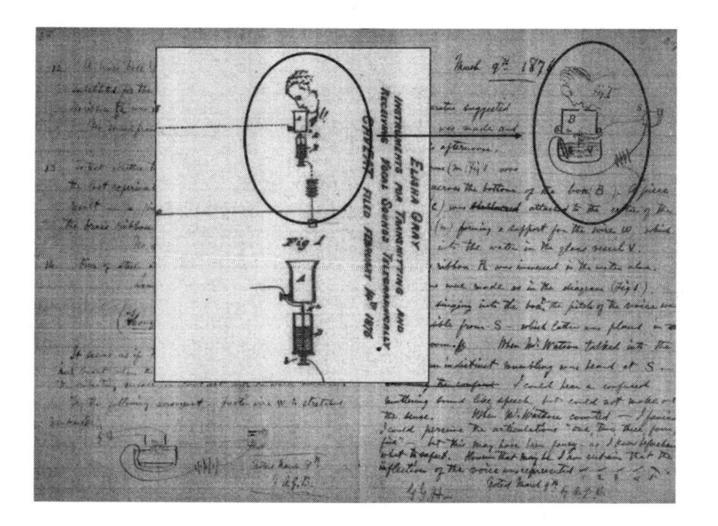

AN AMALGAM OF THE TWO DOCUMENTS HIGHLIGHTING BELL'S PLAGIARISM OF GRAY'S INVENTION—THE OBSERVATION THAT LAUNCHED ME UNEXPECTEDLY INTO A VEXING INTRIGUE.

The drawings left me little room for doubt about where Bell's idea for a liquid transmitter had come from. But, in so doing, they suggested a historical intrigue so at odds with the conventional story of the telephone's invention that I could hardly think where to begin to try to unravel it. I had come to MIT to explore the rivalry between Bell and Edison. But now Thomas Alva would have to wait. I had happened upon a stunning fissure in the polished facade of Bell's legacy; I couldn't help but try to pry the history open from the beginning.

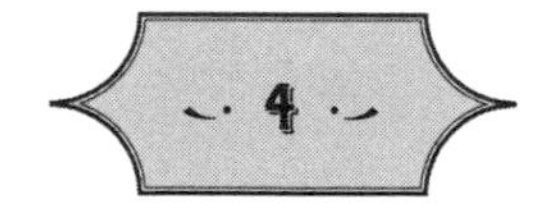

CALLING HOME

ALEXANDER GRAHAM BELL was never one of those mechanically inclined children who excel at taking apart and fixing things. In his later years he told a story, apocryphal or not, that when his father had asked to have his pocketwatch cleaned, young Bell took it apart and washed the pieces with soap and water. As Bell put it, his father was "not enthusiastic over the result."

Bell remained famously clumsy with mechanical devices throughout his life. But he was always exceptionally proficient in the conceptual realm. A bright and dutiful child, he inherited his fascination with speech, sound, and the emerging field of acoustics much as one would inherit a trade. Bell was born in 1847 in Edinburgh, at the beginning of the second decade of the rule of Queen Victoria. Britain was entering an age of industrial expansion. Science and rationality were ascendant. And Bell's family had built the scientific study of speech into a kind of Victorian-era cottage industry.

Bell's paternal grandfather and namesake, Alexander Bell, taught elocution. So did Bell's uncle David Bell, and his father, Alexander

Melville Bell, who published enormously popular texts on the subject and even developed a system called "Visible Speech" that became world-renowned in its day. Visible Speech, in wonderfully grand, Victorian style, attempted to systematically catalog all possible human vocal sounds by assigning each a written symbol that represented the placement of the tongue and lips as the particular sound was uttered.

If the idea sounds vaguely familiar, perhaps it is because George Bernard Shaw, a family acquaintance, immortalized Melville Bell's system in his play *Pygmalion* (the basis of the subsequent musical *My Fair Lady).* Shaw's preface to the play even mentions "the illustrious Alexander Melville Bell, the inventor of Visible Speech." And, coincidentally or not, Shaw gives his memorable character Professor Henry Higgins an address just minutes away from the actual spot on

Bell's father, Alexander Melville Bell.

Harrington Square in London where Grandfather Alexander Bell tutored students.

Shaw's Professor Higgins—as erudite as he is overbearing—surely draws upon the tendency toward grandiosity and stern reverence for science displayed by the real-life patriarchs of the Bell family. Grandfather Bell taught an eclectic assortment of adults with stammers and other speech impediments as well as children whose upwardly mobile families wanted them to improve their elocution. Young Aleck Bell himself was carefully schooled in elocution and high-society etiquette at the hands of both his father and grandfather—not entirely unlike Higgins's pupil Liza Doolittle, Shaw's protagonist in *Pygmalion*.

At the age of fifteen, for instance, the family sent Aleck from their

Alexander Graham Bell in 1863, at age sixteen.

home in Edinburgh to live for a year with his grandfather in London. Bell invariably thereafter called the experience a turning point in his life. During his stay, Aleck worked intensively under his grandfather's tutelage to polish his diction and accent by reading Shakespeare aloud. And elocution lessons were just part of his training. Grandfather Alexander also required him to don a suit jacket and top hat, and even to carry a cane—a teenaged caricature of a dapper Englishman, much like one Shaw himself might have conjured.

Around this time, Melville Bell's system of Visible Speech was attracting interest throughout Britain. It fit the times perfectly, melding a grandiose scientific approach with the appealing, egalitarian prospect of self-betterment. Aleck, as a teenager, became one of its most well-versed practitioners and even took part in his father's frequent lectures. Like a showman's sidekick, Aleck would wait offstage and out of earshot while his father asked members of the audience to suggest difficult or unusual sounds—including words in any language—writing on a chalkboard the distinctive phonetic symbols he had invented for the sounds. Aleck would then return and pronounce sounds he had never heard to illustrate the viability of his father's linguistic scheme.

As Bell later recalled, the symbols at one such lecture called for him to blow a puff of air while the tip of his tongue touched the roof of his mouth. Following his father's written instruction, Aleck made the odd sound and drew a great round of applause in response. A linguist in the audience, Bell later wrote, had suggested to the assembled crowd that the sound—which he called the "Sanskrit cerebral T"—was one of the hardest for an English speaker to utter. Needless to say, the linguist was deeply impressed when Aleck, working only from his father's notations, pronounced it correctly on the first try.

The training in etiquette, elocution, and public speaking that Bell received from his father and grandfather would serve him exceedingly well throughout his life. Much later, for instance, Thomas Watson put Bell's vocal abilities first among the many rewards of having been his assistant, noting:

> *The best thing Bell did for me—spiritually—was to emphasize my love for the music of the speaking voice. He was himself a master of expressive speech. The tones of his voice seemed vividly to color his words. His clear, crisp articulation delighted me and made other men's speech seem uncouth.*

Watson, introverted and tongue-tied prior to his association with Bell, was immensely taken by his colleague's vocal command and flair for public speaking. Long after the telephone brought fame and fortune to both of them, Watson even joined a Shakespearean repertory theater troupe.

Rivaling Bell's vocal gifts was his natural ear for music. One of three brothers, Bell was a sensitive middle child and a gifted musician, studying piano first with his mother and then with a well-known concert pianist, August Benoit Bertini. From the youngest age, Bell could both improvise and play difficult pieces by ear, and he would be captivated by music throughout his life. In his early teens, Bell recalled later, he even dreamt of following Signor Bertini's model and becoming a glamorous, performing musician.

Bell shared his love of music in his close relationship with his mother, Eliza Symonds Bell, a bright and cheerful woman who could hear only with the aid of a Victorian-era speaking horn held up to her ear. Aleck would often play music for his mother as she pressed the horn against the piano's sounding board. The experience was no doubt formative for Bell. Not only would he continue to play the piano for the decades to follow, his concern for the deaf would become a defining and lifelong passion.

ALECK BELL AND his two brothers, Melville (Melly) and Edward, all planned to follow the career path laid out by their father and grandfather. But this latest generation of Bells entered a markedly different field: a renaissance in the science of acoustics was yielding dramatically new ways to study and explore the production, transmission, and perception of sounds. The telegraph had opened up not only a new world of instantaneous telecommunication but also vast new areas of research. Among the major practitioners in this emerging field was Sir Charles Wheatstone, a largely self-taught British scientist whose early work included experiments on the transmission of sound, and who by the mid-1800s held some of England's most lucrative telegraph patents.

When Bell was sixteen years old, his father took him along on a visit to Wheatstone's laboratory in London. Sir Charles had, some years earlier, constructed an artificial speaking machine, a feat of considerable interest to Melville Bell given his work in elocution. In fact, Wheatstone had rebuilt his own version of a contraption invented decades earlier by an eccentric researcher named Baron Wolfgang von Kempelen from Hungary, who set out to create the machine based upon his pioneering insight that the sound of human speech must consist of nothing more than vibrating air.

Wheatstone's tabletop device had a bellows that blew air into a box filled with an assortment of pipes with holes, valves, and levers that could be adjusted by hand. With practice, Wheatstone explained, a user could manipulate the resonating cavity to simulate the sounds of different vowels and consonants. Aleck was fascinated. As he later wrote:

> *I saw Sir Charles manipulate the machine and heard it speak, and although the articulation was disappointingly crude, it made a great impression upon my mind.*

Back in Edinburgh, Melville capitalized on Aleck's enthusiasm by challenging him and his older brother Melly to construct their own "talking machine." The boys had great enthusiasm, but the task was not easy. After an arduous effort and much experimentation, the brothers came up with a workable prototype of rubber, wood, and wire, replete with a bellows from a parlor organ to drive air into the machine and a small keyboard to control its parts.

Bell recalled that, while he and Melly were repeatedly discouraged in their quest to make the machine speak, the project taught him the value of persistence. As he wrote over four decades later, in 1909:

> *The making of this talking-machine certainly marked an important point in my career. It made me familiar with the functions of the vocal cords, and started me along the path that led to the telephone.*

Bell acknowledged, though, that the two teenagers had more on their minds than the advancement of scientific knowledge. After assembling the prototype, the boys snuck the machine out to the common stairwell of their town house in Edinburgh and "made it yell." As Bell later recalled, the noise that issued from the contraption sounded remarkably like the plaintive wail of an infant:

> *We heard someone above say, "Good gracious, what can be the matter with that baby" and then footsteps were heard. This, of course, was just what we wanted. We quietly slipped into our house, and closed the door, leaving our neighbors to pursue their fruitless quest for the baby. Our triumph and happiness were complete.*

From Bell's rich childhood emerged a young man whose talents and interests matched the tide of technological change around him remarkably well. He just didn't know it yet.

IN THE FALL of 1863, Bell—a sixteen-year-old riding high on his family's reputation and fresh from his grandfather's intensive tutoring—took his first full-time job, teaching music and elocution at Weston House Academy, a boarding school for boys in the town of Elgin on Scotland's northeastern coast. The job gave him an opportunity to draw upon many of the powerful influences on his life: his training in elocution, his love of music, and his exposure to the science of acoustics. Bell thrived in the position. After two years at the job, the young man used his spare time to try to make his own mark on the young science of acoustics. At the age of eighteen, in his first-ever professional experiment, Bell focused on a topic that would hold tremendous import for him in the years to come: the idea that sound waves could induce sympathetic vibrations.

In his experiment, Bell placed a tuning fork in front of his open mouth. He made vowel sounds and noted that the tuning fork would resonate more loudly when he held his tongue in some positions than it would in others. Bell experimented further with what we now call overtones, trying to measure the precise pitches of varying vowel sounds and, in so doing, seeking systematically to understand the effects he observed.

With his father's encouragement, Bell sent a report of his experiments to Alexander John Ellis, one of Britain's leading authorities on phonetics. Ellis replied that Bell's research repeated the work of the German scientist Hermann von Helmholtz, who had similarly sought to detect the musical tones of vowel sounds with tuning forks. Helmholtz, a physicist and medical doctor, was a preeminent figure in the field of acoustics who would, from a top position at the University of Berlin, become well known throughout Europe.

It was, of course, an impressive feat for an eighteen-year-old to unknowingly repeat Helmholtz's research. Ellis headed the London Philological Society, a scholarly group concerned with the study of speech and language, precursor to the modern-day field of linguistics. Based on what he had seen of young Bell's achievements, he nominated Bell for membership. Even more important, Ellis encouraged Aleck to continue his research and lent him Helmholtz's pathbreaking work, *On the Sensations of Tone as a Physiological Basis for the Theory of Music.* In the book, Helmholtz describes his invention of a machine called a "tuning fork sounder," a device built upon much the same line

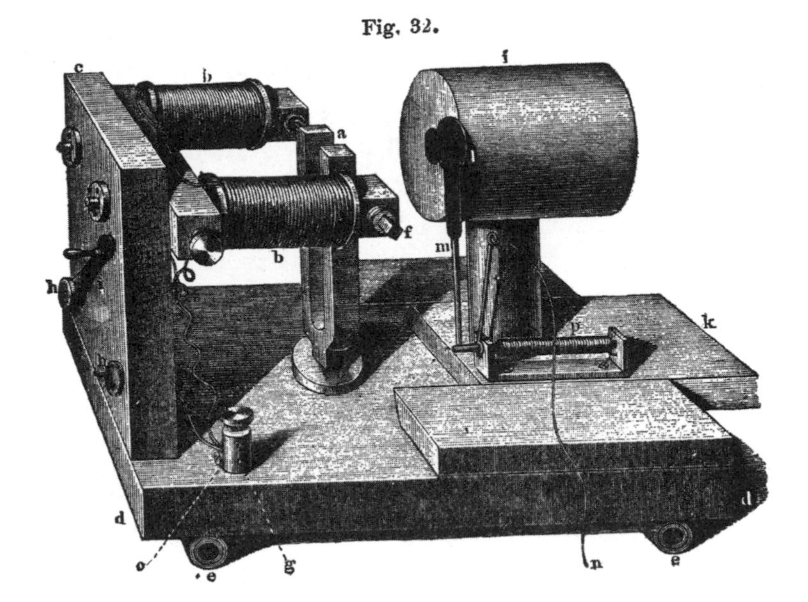

HERMANN VON HELMHOLTZ'S "TUNING FORK SOUNDER" DEVICE, CIRCA 1865.

of thinking as Bell's tuning fork experiment and, in essence, a far more sophisticated version of Baron von Kempelen's speaking machine. Helmholtz's device used an intermittent electric current to keep a tuning fork in constant vibration. The tuning fork stood before a resonator cavity made of a cardboard tube fitted with a movable cover. By opening and closing the opening to the tube, the device could create remarkably human vowel sounds.

At the time, Helmholtz's work had yet to be translated into English (Ellis himself would eventually do the job) and Aleck couldn't read German. Somehow, though, either from Helmholtz's diagrams or Ellis's explanations, Bell arrived at the completely erroneous conclusion that Helmholtz had not only managed to modulate vowel sounds but *to transmit them along telegraph wires*. The mistaken notion thoroughly fascinated Bell and spurred him to experiment with electricity.

Soon after hearing of Helmholtz's work, Bell read up on the subject of batteries and built a small telegraph of his own.

Perhaps the best account of this confusion over Helmholtz's work is relayed, much after the fact, by Bell himself via his longtime assistant Catherine MacKenzie. As MacKenzie recounts in the first biography of Bell, *Alexander Graham Bell: The Man Who Contracted Space* (1928):

> *"I thought that Helmholtz had done it," Bell used to say, "and that my failure was due only to my ignorance of electricity. It was a very valuable blunder. It gave me confidence. If I had been able to read German in those days, I might never have commenced my experiments!"*

In Bell's estimation, then, his telephone research was spawned by a colossal misunderstanding. According to MacKenzie, Bell didn't recognize his mistake until he was about to leave Great Britain in 1870 and finally read a copy of the French edition of Helmholtz's work. Only then did Bell learn he had been trying to reproduce a feat Helmholtz himself had never actually accomplished.

Mistaken or not, there is little doubt Bell's notions about Helmholtz's work were of signal importance to the development of his thinking. By the time Bell headed to America, as MacKenzie notes, the idea of sending sounds over the telegraph had "rooted itself firmly" in his imagination.

THE BELL FAMILY'S precipitous departure from Great Britain in 1870 came as a result of tragedy. Grandfather Alexander had died in 1865, and in the several following years, both of Bell's brothers died of tuberculosis. Bell's younger brother Edward succumbed to the disease in 1867, when he was just nineteen years old. When Melly died in 1870, at age twenty-five, his heartbroken parents decided to emigrate to North America to make a fresh start.

The move came at a difficult time for Aleck Graham Bell. He had begun to make a name for himself and hoped to complete his studies. In 1868, he had passed his entrance exams and matriculated as a student at the University of London. He had also begun to use his understanding of the mechanics of elocution to teach deaf children to speak, an effort that had met with remarkable success and won him growing recognition.

But Bell's parents were certainly justified in worrying about the health of their surviving son. Throughout his early life, Aleck had suffered from chronic maladies and, as a young man, had repeatedly worked himself into a state of exhaustion. He would frequently complain of headaches severe enough to make reading impossible. Bell's parents felt it imperative to find a healthier environment for him.

They also knew that Melville Bell's reputation, though flagging in England, was ascendant in North America after word of the Visible Speech system had crossed the ocean. On a visit to Boston in 1868, for instance, Melville had given the prestigious Lowell Lectures to widespread acclaim; according to Melville's own proud reports home, Harvard University president Thomas Hill was among the enthusiastic fans who had personally taught himself Visible Speech from Bell's book.

Unmoored by the deaths in the family, and in the hopes of protecting their remaining son's health, Bell's parents chose to move to Brantford, Ontario, where family friends from Scotland had taken up residence. Aleck joined them, but as it turned out, in those years, at least, he would spend little time in Canada. Rather, trading once again on his father's reputation, twenty-four-year-old Aleck Bell would find his fortune in Boston, teaching the deaf and lecturing at Boston University.

NO ANSWER

I WISH I COULD say that I jumped on the telephone case with the zeal of Bell's literary contemporary Sherlock Holmes. But I did not. After eagerly making a photocopy of Gray's caveat, I returned to my office and—I remember this part quite clearly—placed the document *underneath* a pile of papers on my desk. It sat there for over a week, buried physically and figuratively as I weighed it against the information contained in the many historical texts and biographies I had begun to collect about Bell.

At first, I think, I set the information aside because I didn't want to learn something untoward about Bell. After all, I had been drawn to write about him because I admired him, not just as a creative inventor but as a humanitarian. The seemingly incriminating connection I had discovered was so at odds with the story I had planned to tell that I simply didn't know what to make of it. To be sure, I had stumbled upon a tantalizing find. If it panned out, the accepted, almost ubiquitous historical tale of the telephone's invention could be turned on its head. But, for better or worse, I wasn't sure how to proceed. I didn't know who would believe me or how I could prove it even if Bell *had* stolen the design for the telephone. How could a journalist hope to set straight such a high-profile historical record more than a century after the fact? Myths, erroneous or not, are powerful.

I knew that accusing Alexander Graham Bell of malfeasance or worse would mean embarking on a long, probably frustrating venture. It would mean confronting the failings of the U.S. legal system, which had heard and rejected numerous challenges to Bell's claim to the telephone. It would mean questioning the work of generations of trained and respected historians. It would mean defying stacks of texts and reference books that credit Bell with the telephone's invention. The prospect seemed more than a little daunting.

When I did start to grapple with the information I had found, the major secondary sources I consulted about Bell only made matters worse. Perhaps none of these was more confounding than Robert Bruce's highly regarded 1973 biography, *Bell: Alexander Graham Bell and the Conquest of Solitude.* Bruce, a Pulitzer Prize–winning historian, goes so far as to cast the aspersion that Elisha Gray stole Bell's idea. "Was it entirely a coincidence" that Gray filed his claim at the same time as Bell did, Bruce asks. As he writes:

> *If Gray had prevailed in the end, Bell and his partners, along with fanciers of the underdog, would have suspected chicanery. After all, Gray did not put his concept on paper nor even mention it to anyone until he had spent nearly a month in Washington making frequent visits to the Patent Office, and until Bell's notarized specifications had for several days been the admiration of at least some of "the people in the Patent Office."*

Bruce's tone is so assured and seemingly authoritative on the subject of Bell's claim that it shook my confidence deeply. I thought I must be mistaken. For some time as I puzzled over the case, I even overlooked the obvious point that Gray's pathbreaking design for a liquid transmitter is nowhere depicted in Bell's patent. Even stretching to give Bruce the benefit of the doubt, Bell's "notarized specifications" make only the most oblique and passing mention of such a possibility. Meanwhile, the liquid transmitter makes its first appearance in Bell's notes—in a form virtually identical to Gray's design—some three weeks *after* Gray filed his caveat. And Bell's notebook offers no indication that he had ever experimented with such a transmitter prior to that time, *nor had he fully succeeded in transmitting intelligible speech with any other method.* Authoritative-sounding or not, Bruce's conjecture simply does not fit the facts.

But what *were* the facts? Could I hope to discern them after so much time had passed?

One aspect of the question tantalized me in particular. Alexander Graham Bell conducted his telephone research less than two miles from my office. He sought advice from MIT professors whose archival papers still resided nearby on campus. Bell made the world's first public demonstration of the telephone in 1876 at the American Academy of Arts and Sciences, which was, at the time, located directly across the river in downtown Boston. And, in October 1876, when Bell and Watson first tested their machine "long distance" over a two-mile stretch of telegraph wire from Boston to Cambridgeport, Watson had sat in a warehouse that, as near as I could tell, once occupied the parking lot outside my building. The geographic proximity of the whole story made it seem maddeningly within reach, like a treasure buried somewhere beneath my feet.

It was a beautiful autumn in Boston, with striking fall foliage in oranges and reds. One especially sunny afternoon, about a week after finding Gray's caveat, I put aside the stacks of biographical material and set out on a long, quixotic walk to learn more about Bell by retracing his steps in Boston. What I learned instead is that history's physical traces can be infuriatingly ephemeral. I walked for many hours to parts of Boston I had rarely if ever visited. But I found little remaining evidence of the major landmarks of Bell's life.

I headed first for Bell's workshop at 5 Exeter Place, only to learn that it had long since fallen victim to urban renewal. A painting by W. A. Rogers, frequently reprinted in the secondary literature about Bell, depicts his workshop as it stood in March 1877. What comes across perhaps most notably is how dowdy, Victorian, and *ordinary* it was, with its coal stove, bare wood floor, and patterned wallpaper.

Painting by W. A. Rogers from a sketch in March 1877 depicting Thomas Watson and Alexander Graham Bell in their workshop at 5 Exeter Place in Boston.

Today, however, the boardinghouse where Bell worked is hard to even imagine amid the city's towering office buildings and the ongoing construction of Boston's "Big Dig." The only remnant of its former existence is a forlorn bronze plaque placed by a local historical society to commemorate the fortieth anniversary of the telephone's invention. Standing largely unnoticed near a sea of traffic on Boston's central artery, the quaint and formal plaque reads:

> *Here Alexander Graham Bell transmitted to Thomas Augustus Watson the first complete and intelligible sentence by telephone, March 10, 1876.*

Making my way through narrow side streets to the heart of the city, I tried next to locate the Charles Williams machine shop where almost all of Bell's earliest telephone devices were constructed. In Bell's day, Williams's shop—employing roughly two dozen machinists—had buzzed and clattered with the din of lathes and metalworking tools as workers turned out prototypes for an array of strange new electrical devices, from telegraph relays to galvanometers.

As I soon learned, no trace of the shop remains today. Not only has the building been demolished, its address, 109 Court Street, no longer even exists. In the 1960s, the city shortened Court Street to make room for a new City Hall at Government Center. Today, not even a plaque marks the site of the Williams shop—once the world's epicenter of telephone research and home for a time to Bell's personal workshop. Instead, in one of history's ironies, the spot, near as I could tell, is now occupied by an increasingly rare cluster of outdoor pay phones.

Finally, on Beacon Hill, near Boston's statehouse with its magnificent Bulfinch dome, I found one of the few important sites from Bell's time that has survived: the historic Boston Athenaeum Building, housing one of the nation's oldest membership libraries. Here, Bell first presented his telephone research to the world before a gathering of the American Academy of Arts and Sciences. Built in 1847, the year Bell was born, the granite building at 10½ Beacon Street still exudes the charm of his era, outside and in.

I talked my way past the guard at the entrance to catch a glimpse of the main hall on the first floor, a grand room ringed by a small balcony and large, paned windows along the back wall. There, inside the hushed, historic library, I found the tangible connection to Bell I had set out in search of: the Athenaeum, with its unmistakable Boston Brahmin pedigree, remains all but unchanged from Bell's day.

THE CHARLES WILLIAMS MACHINE SHOP IN BOSTON, CIRCA 1870.

As I headed back through the front lobby to the spare, historic front entrance, it was easy to conjure the scene of Bell's 1876 presentation. I imagined horse-drawn carriages clattering down the cobblestone-lined street outside and pulling up in front of the building. I imagined formally attired gentlemen in top hats earnestly popping out onto the curb and streaming in to the Academy's meeting, curious to hear from a young scientist presenting what he called his "researches in telephony." But once back outside, the press of the city's bustling traffic quickly intruded on my reverie.

PERHAPS THE MOST compelling portrait of young Aleck Bell's life in Boston comes from his assistant, Thomas A. Watson. In 1874, when Watson first met him, Bell was twenty-seven years old and thriving. Since arriving in Boston several years earlier, Bell had served as a most successful ambassador for his father's system of Visible Speech. With Bell's own natural empathy for his students and his gifts as an instructor, he had also distinguished himself by teaching a number of deaf children to talk. By using pictures based upon his father's system, Bell found he could illustrate for the children exactly how to position their mouths and tongues in order to speak correctly. The results were dramatic. Before long, dozens of leading educators from throughout the region sought to learn more about Bell's techniques, and newspapers chronicled the success of his School of Vocal Physiology. As a result, Bell soon built upon his core of a dozen private pupils—many of whom were from wealthy families—to take on additional duties as a professor of "vocal physiology and elocution" at Boston University, which had just opened in 1869. As he had in Scotland, Bell did his inventing in his spare time.

At the time, after having lived in the city for several years, Bell had accepted an offer from the wealthy family of one of his students—young George Sanders—for free room and board at their stately home in Salem, Massachusetts, in exchange for his tutoring services.

Bell commuted into Boston by train, and initially set up a room for his acoustic experiments at the Sanderses' home. Thomas Sanders soon agreed to help underwrite Bell's experiments in exchange for a share of the profits should the work prove commercially viable. Before long, Bell would make an additional, similar arrangement with another parent of a deaf student, the wealthy and worldly Gardiner Greene Hubbard.

Before meeting Bell, Watson had been an enterprising, earnest teenager with little formal education, who had worked for a living since he

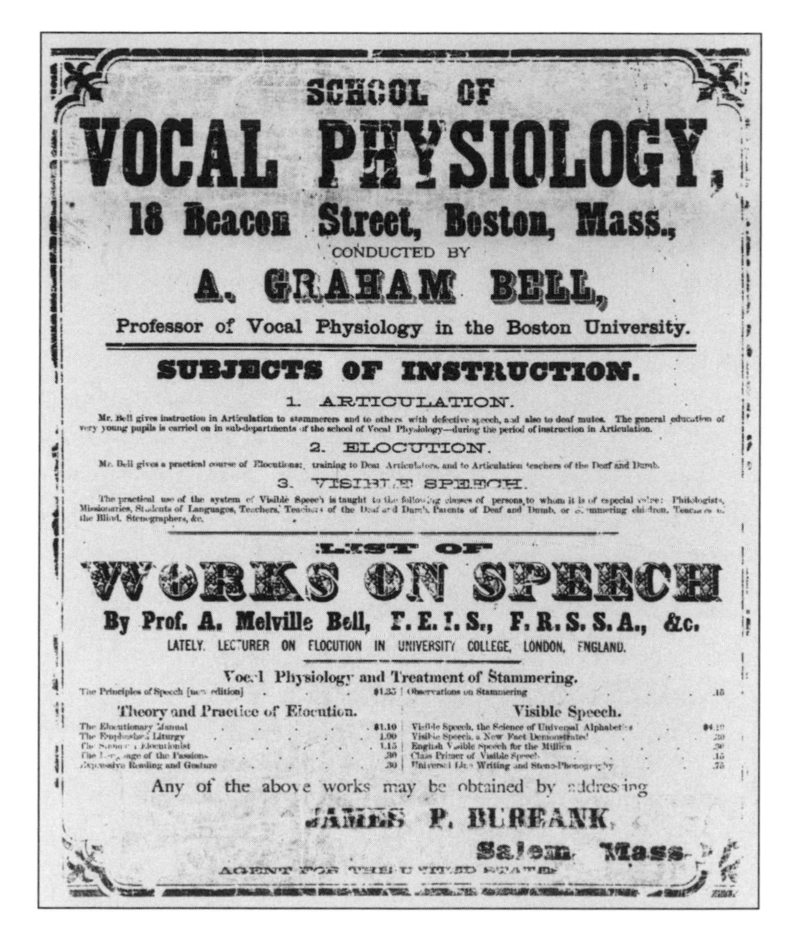

An advertisement for Alexander Graham Bell's School of Vocal Physiology on Beacon Hill in Boston, circa 1874.

was thirteen years old. In July 1872, at the age of eighteen, Watson landed a job at the Charles Williams machine shop on Court Street. As Watson recalls, it was a thrilling place for a young man like him—a hub for visionary inventors attempting to build all sorts of machines that tapped the potential of the exciting and little-understood "power of electricity."

Among these young inventors was Thomas Edison, who, in the late 1860s, set up his office in the same building as the Williams shop to best avail himself of its services. It was here, in fact, that Edison won his first patent—for an electrical vote recorder—only to realize that no one would buy the machine. Perhaps, given today's controversies over voting technology, Edison's invention simply came a century and a half ahead of its time. Nonetheless, it was, Edison said later, the last time he would invent anything without first making sure there was a market for it.

As a machinist, Watson's job was to make prototypes to the specifications of the shop's patrons, including people like Edison and Moses Farmer, another respected electrical researcher of the day. As Watson memorably recalls in his autobiography, *Exploring Life* (1926), no one in the Williams shop ever knew what to expect. Watson certainly did not expect the dramatic arrival of a man who would change his life:

> *One day early in 1874 when I was hard at work for Mr. Farmer on his apparatus for exploding submarine mines by electricity and wondering what was coming next, there came rushing out of the office door and through the shop to my workbench a tall, slender, quick-motioned young man with a pale face, black side-whiskers and drooping mustache, big nose and high, sloping forehead crowned with bushy jet-black hair. It was Alexander Graham Bell, a young professor in Boston University, whom I then saw for the first time.*

Bell stormed into the shop holding two small instruments Watson had crafted for him. Breaking with normal procedure, he headed straight onto the shop floor to complain directly to Watson that the machines had not been built according to his instructions. Bell's demeanor was exceedingly formal; but he was also frequently hot-tempered, and in

THOMAS A. WATSON IN 1874, THE YEAR HE FIRST MET ALEXANDER GRAHAM BELL.

this case, he got right to the point, demanding that Watson correct his mistakes. Watson was happy to comply. He listened with interest to Bell's explanations about the strange contraptions he had constructed with no idea what they were intended for.

The pair of instruments Watson pledged to rebuild descended directly from Bell's tuning fork experiments in Scotland. Since moving to Boston, Bell believed he had found a practical application for his ideas about sympathetic vibration. He knew the telegraph industry was having difficulty keeping pace with the voluminous number of telegrams being sent. More and more unsightly wires were rapidly being strung on telegraph poles that, at considerable expense to the industry, were proliferating across the continent. In parts of some cities like Boston, the tangle of overhead telegraph wires was becoming oppressive, all but blocking out the sky. As a result, Western Union had announced that it was willing to pay up to $1 million to the inventor who could ease the congestion by allowing telegraph wires to carry multiple messages simultaneously.

One researcher and patron of the Williams shop named Joseph Stearns had recently invented a "duplex telegraph" that allowed a single telegraph wire to carry both an outgoing and an incoming message simultaneously. Stearn's scheme used a parallel circuit at each end of the main telegraph line that would blot out only the outgoing messages from that end of the line, leaving the telegraph device free to accurately receive incoming messages at the same time. It was an important advance. But Bell thought he had an even better solution with what he called his "harmonic, multiple telegraph."

Bell's concept was simple. Using the principle of sympathetic vibration, he hoped to develop a paired transmitter and receiver that could be tuned to a particular pitch. That way, it might be possible to send multiple messages—at different pitches—over a wire simultaneously. In Bell's conception, the telegraph receivers would only vibrate in concert with the messages sent by the transmitters they were tuned to.

Bell had asked Watson to build an early prototype of the multiple telegraph design. The transmitter used a "reed" made out of a small, thin strip of steel that was mounted above an electromagnet with an adjustable contact screw like that found on an electric buzzer. When Bell attached the transmitter to a battery, the steel reed would vibrate and emit a sound; by moving the reed in and out of contact with the electromagnet, he could create an intermittent current that corresponded to the pitch of the reed. According to Bell's plan, this vibrating, intermittent current would pass through the telegraph wire and set a reed in a distant receiver into sympathetic vibration—just as the

By the 1870s, tangles of telegraph wires were becoming a blight on some urban areas. The industry sought devices that could send multiple messages on a wire simultaneously.

sound waves from Bell's voice had vibrated a tuning fork held in front of his mouth. It was a line of thinking that would, before long, lead Bell directly to what we now know as the telephone.

AFTER WANDERING ALL afternoon in search of Bell's history in Boston, I was exhausted when I made my way back to MIT. With little tangible evidence to hang on to, I wondered gloomily whether historical questions like mine could ever be answered definitively. It was already evening by the time I reached the long, carpeted corridor to my office.

The building was quiet, but the door to the office next to mine was open and the light was on. Inside, David Cahan sat engrossed before his computer. Cahan, a friendly man with a big midwestern smile and slightly stooped shoulders that made him seem at once warm and professorial, taught the history of science at the University of Nebraska. He was also, as fate would have it, one of the world's leading scholars on the life and work of Hermann von Helmholtz. For nearly fifteen years, Cahan had been working on a definitive biography of Helmholtz, and his office shelves, piled with stacks of notes and manuscript pages, reflected his long labors. He glanced up and greeted me warmly, presumably happy not to be the only one still toiling.

Once in my office, I sunk into my desk chair and pulled out my notebook. I couldn't help but think of my neighbor next door. I thought of all the historical mysteries that must surely have flummoxed Cahan in his career. And yet there he sat, undaunted even after long years of doggedly sifting fact from fiction about Helmholtz's life. I wasn't confident that I could uncover enough detail about what had transpired at the birth of the telephone to explain how Gray's secret design seemed to have wound up in Bell's notebook. But, for the year at least, I realized I was literally surrounded by historians brimming with expertise, people like Cahan who were devoting their lives to studying and understanding the history of science and technology. There was no question about it: I would just have to enlist their help to unearth what I could about the real story behind the telephone's invention.

OPERATOR ASSISTANCE

ON A LATE OCTOBER afternoon in 1874, Alexander Graham Bell paid a fateful visit to the Hubbard residence in Cambridge's wealthiest neighborhood. For the past year, Bell had been tutoring Mabel, the Hubbards' sweet and vivacious sixteen-year-old daughter, who had been deaf since a bout of scarlet fever when she was five years old. Now Bell had been invited to join the family for tea. As he approached the house at 146 Brattle Street, Bell paused to admire its grandeur. The impressive Italianate mansion overlooked the Charles River. Surrounded by formal gardens, a stable, and a greenhouse, it stood just down the road from Henry Wadsworth Longfellow's equally luxurious Colonial-style estate.

Since arriving in Boston, Bell had traveled in some fancy circles. The well-heeled Sanders family, for instance, had included him in many social gatherings. They lived in a large house dating from the Colonial period topped with a Captain's Walk that overlooked the Atlantic. But the tasteful grandeur of the Hubbard residence was striking by almost any comparison. Bell had already noticed how beautifully dressed, supremely well mannered, and bright Mabel was. She was easily the most winning student he had ever tutored. Mabel's home even more clearly reflected her family's wealth and upper-class pedigree.

Mabel's father, Gardiner Greene Hubbard, an attorney and entrepreneur, had offices in both Boston and Washington. His father, Samuel Hubbard, had sat on the Massachusetts Supreme Court. His maternal grandfather and namesake, Gardiner Greene, was reputed in his day to be the wealthiest man in Boston. And the Hubbards' Boston roots went deep. For example, Hubbard attended Harvard Law School in the early 1840s; his first American ancestor, William Hubbard, a notable Massachusetts pastor, had graduated from Harvard College some two hundred years earlier, in 1642.

Bell had already met Gardiner Hubbard professionally in conjunction with Hubbard's impressive work on behalf of the deaf. After Mabel lost her hearing, Hubbard refused to relegate one of his four charming daughters to an asylum, as was the custom of the day. Instead, he devoted himself to creating opportunities not only for Mabel but for other deaf children as well. Among his accomplishments in this regard, Hubbard was a founder and first president of the Clarke School for the Deaf in Northampton, Massachusetts—an institution that would gain world renown for its innovative methods of teaching the deaf to speak and helping them integrate into the hearing world.

Thanks to Hubbard's efforts and his daughter's natural talent, Mabel became an accomplished lip reader, and her comprehension was impeccable. By the age of twelve, she was attending classes in a regular school with hearing children two and three years her senior. Nonetheless, Hubbard felt her speech could be improved. In August 1873, he sought out Bell's services shortly after his daughter's return from an extended sojourn in Europe with her mother and sisters.

Now, some fourteen months later, Bell drank up the hospitality and attention of the Hubbard family. Mabel's mother, who fully rivaled her husband in her energetic nature and rarefied background, charmed him from the start. Gertrude McCurdy Hubbard was the daughter of an established and well-to-do New York family. Her father, Robert Henry McCurdy, was a trustee and founder of the vast Mutual Life Insurance Company.

Not only did Gertrude Hubbard oversee a busy residence and social schedule in Cambridge, New York, and Washington; she was also highly educated and self-directed. When Mabel was small, for instance, she took it upon herself to learn Hebrew so she could read the Old Testament in its original language. More recently, when Gertrude Hubbard traveled with her family to Europe, her husband returned home after several months to tend to business. Mrs. Hubbard, however, remained abroad for two more years to single-handedly tour her four daughters through Geneva, Vienna, Rome, Florence, Paris, and London.

In the Hubbards' formal drawing room, which was decorated in High Victorian style, with red velvet wallpaper, gilded drapes, and gaslights with crystal fixtures, Bell entertained his hosts on the family's grand piano.

After playing a favorite sonata, Bell seized the opportunity to tell Gardiner Hubbard about his telegraphic research. As the two men stood together by the piano, Bell gave Hubbard a demonstration that stemmed directly from his original experiments with tuning forks in Scotland: he showed how, by stepping on the sustain pedal and bending over its open top, he could make the instrument's undamped strings vibrate sympathetically with whatever note he sang. Sure enough, Hubbard saw, the piano strings echoed the precise pitch of Bell's singing.

Bell explained that the effect was part of his research. Instead of using air, he could get the same effect by carrying the vibrations over a wire. That way, he should be able to make a string or reed sound from hundreds of miles away. Using the technique, Bell said, he believed he could build a multiple telegraph capable of sending six, eight, or even more messages over a single telegraph wire simultaneously.

Bell's report visibly and uncharacteristically excited his host. Hubbard was a serious, formal patrician, who invariably wore his dark suit buttoned and his graying beard long and untrimmed. "I brought the subject before the Hon. Gardiner G. Hubbard," Bell wrote home to his family soon after his visit.

> *I explained the system to him in confidence and was surprised at the way in which he, an undemonstrative man, received the explanation. He called his wife and made me explain it all to her, and when she raised some objection to one point, he answered it himself saying "don't you see there is only one* air *and so there need be but one* wire!"

From earlier letters home, it is clear that Bell knew of Gardiner Hubbard's interest in the telegraph long before visiting the Hubbard household. A letter in March 1874, for instance—months before Bell's

GARDINER GREENE HUBBARD.

visit—mentions the so-called Hubbard Bill introduced in the U.S. Congress to dismantle the Western Union monopoly and nationalize the telegraph industry. Bell tells his parents about tutoring Mabel and muses on the possibility of writing to her father about his telegraphic research given Hubbard's high-profile position in the telegraph business.

Despite whatever plans Bell may have laid in advance, though, he could not have expected the swiftness with which his work would capture Hubbard's entrepreneurial imagination. Hubbard instantly recognized the potential of the invention Bell outlined. That October day in Hubbard's elegant parlor, Bell's demonstration set in motion an extraordinary chain of events. Almost immediately Hubbard began to use his own wealth, connections, legal expertise, and political savvy to shape Bell's fledgling telegraphic research.

Bell wrote home on October 20, 1874, after his visit to the Hubbards:

> *I am tonight a happy man. Success seems to meet me on every hand. . . . After my last interview with [Hubbard] he had gone down to Washington and searched the Patent Office to find whether my idea had been taken up by anyone else—and he now offers to provide me with* funds *for the purpose of experimenting if we go into the scheme as partners.*

AS A LAWYER specializing in patents, Gardiner Hubbard was naturally interested in inventions. In his legal practice he had helped to secure patent protection for everything from new machines used in the manufacture of shoes to specialized saws for milling lumber. But Hubbard had his own reasons for so quickly and enthusiastically backing Bell's research: he had become intimately involved with the politics of the U.S. government's efforts to regulate the lucrative and growing telegraph industry.

For at least six years before Bell's visit, Hubbard had devoted much time and energy to a campaign calling for the government to assume control of the telegraph industry and bring it under the purview of the U.S. Post Office. In 1868, at the request of the Postmaster General, Hubbard had compiled a report on the postal-telegraph systems in other countries, most of which were under government control. In the countries Hubbard reviewed, the growth of the telegraph infrastructure and escalating telegram volume had led to reduced prices. But Hubbard's analysis showed that in the United States, Western Union, with its near monopoly control of the telegraph lines, had frequently raised its rates despite the growth of its business.

In a widely read article in the *Atlantic Monthly* on the subject, Hubbard wrote that the population of Britain spent some $5 million to send roughly 18 million telegrams. In the United States, customers spent close to twice that much to send just 13 million telegrams. The size of the country, Hubbard believed, could not fully account for the difference. Rather, he contended, Western Union was overcharging its customers and thereby hurting the U.S. economy.

Congress debated the merits of the Hubbard Bill on several occasions. The plan was intriguing. Hubbard did not call for the U.S. Post Office to actually take over the telegraph industry. Instead, he proposed that Congress authorize the capital to create a private corporation—to be known as the United States Postal Telegraph Company—that would enter into a contract with the Post Office. This private corporation would build and oversee a new telegraph network in the service of the federal government. And perhaps most notable of all: it would be run by a consortium led by Hubbard and his business associates.

In Congress, Hubbard's proposal caused controversy. Some representatives viewed the plan as the work of a public-spirited patriot; others saw it as the scheme of an overzealous entrepreneur. Either way, in his efforts to persuade Congress, Hubbard was repeatedly outmaneuvered by William Orton, the hard-driving and politically well connected head of Western Union. For instance, around the time of the debate over the Hubbard Bill, Orton offered all U.S. congressmembers the equivalent of the "franking" privileges offered by the Post Office, issuing them a card with which they could walk into any telegraph office and send an unlimited number of messages for free. In the face of such political tactics, the Hubbard Bill never gained nearly enough political traction to be enacted.

Of course, Orton's tactics were relatively tame by the standards of the day. Political corruption would reach a kind of high-water mark in the coming 1876 presidential election between Republican Rutherford B. Hayes and his Democratic challenger Samuel Tilden, the governor of New York. In that remarkable contest, Tilden, running as a reformer who had broken up the corrupt, New York–based political machine of "Boss" Tweed, would win the popular vote by a considerable margin. But controversy over fraud and intimidation in Florida and several other southern states would push the nation to the brink of a constitutional crisis: a special electoral commission with a one-vote Republican majority would throw out enough contested Tilden ballots to hand Hayes a slim and highly questionable Electoral College victory.

Hubbard's experience in the midst of a political climate pervaded by patronage and corruption had only deepened his antipathy toward Western Union. Little wonder, then, that when Bell unexpectedly outlined an invention with the promise of revolutionizing the telegraph industry, Hubbard immediately saw an opportunity. With Bell's multiple telegraph, Hubbard reasoned, he could either return to Congress to champion a telegraph network built around Bell's efficient new technology, or he could use patent protection to found a telegraph company himself. Bell's kernel of an idea for a harmonic, multiple telegraph, Hubbard figured, might—just possibly—allow him to launch a new telecommunications network to rival Western Union after all.

THE MORE I learned about Hubbard's role in Bell's early efforts to commercialize his telegraph research, the more intrigued I became. For one thing, I was struck by how successful Hubbard had been with several similar ventures closer to home.

Thanks largely to Hubbard's energy and resources, for example, the city of Cambridge installed gas streetlights and a system to deliver clean water years before any other municipality in the Boston area. Hubbard helped to organize the Cambridge Gas company in 1853 and the Cambridge Waterworks two years later, serving as the latter's first president. Even more remarkably, Hubbard helped create a horse-drawn trolley system between Cambridge and Boston. With Hubbard's active backing, the so-called Cambridge Railroad Company was both operational and popular by 1856, disproving the many people who had argued that it would never attract enough passengers to be viable. It was the first municipal trolley of its kind constructed anywhere in the United States outside of New York City, an astonishing feat for a town with just 15,000 residents.

These successful ventures shared an important feature: each combined a heavy dose of public-spirited action with unvarnished self-interest. Although Hubbard had inherited a fortune, he had lost a good deal of it in the 1840s speculating on wheat. Nonetheless, he still had his legal practice, and, even more important, he owned a great deal of prime Cambridge real estate, much of which he would eventually develop into single-family homes. Providing municipal services that made Cambridge a more desirable place to live improved life for Hubbard and his neighbors—and bolstered the potential return from his real estate holdings.

In fact, Hubbard's reputation as a "public man" in these kinds of endeavors was such that Thomas Sanders, Bell's benefactor, landlord, and mentor, was concerned when he first learned that Bell had told Hubbard of his research. Sanders, of course, had already been offering a modest amount of financial support for Bell's work and he wasn't sure he trusted Hubbard. As Bell wrote to his family,

> *Mr. Thomas Sanders said he thought it was unwise to have told the idea to Mr. Hubbard. Public men are so corrupt and the idea*

will be worth thousands of dollars to Mr. Hubbard if he succeeds in buying out the companies. Whereas, if the prior companies were to use the plan—the effect would be to raise the value of all the lines and the government would have to pay more in order to buy them out. Altogether Mr. Sanders thought I should at once protect myself.

At Sanders's urging, he and Bell visited a Boston patent lawyer named Joseph Adams. Their plan was to file a caveat with the U.S. Patent Office on Bell's idea for a multiple telegraph. As Bell tried to perfect a workable prototype, they hoped the caveat would protect his rights to the invention he envisioned. Sanders agreed to pay the legal expenses just as he had so far provided funds for Bell's research.

Despite Sanders's reservations and his push for legal protection, he also realized that Hubbard could be a great asset to the success of Bell's endeavors. Hubbard quickly came in on the deal, and the three men formed a team. They decided to divide any profits from the invention three ways, with Sanders and Hubbard sharing the cost of Bell's experiments and legal fees. In addition, the investors offered to pay for an assistant to help with the experiments, paving the way for Bell's close association with Thomas Watson.

Hubbard's connections, energy, and savvy were on display from the start. He convinced the team to switch to his powerful associates at the Washington, D.C.–based law firm of Pollok & Bailey for legal advice. Anthony Pollok, one of the most politically powerful attorneys of his day, had worked with Hubbard on many projects. Among these, he had been chosen by Hubbard as part of the consortium outlined in the Hubbard Bill to oversee the proposed United States Postal Telegraph Company.

With Pollok's help, Hubbard also immediately convinced Sanders and Bell to withdraw their efforts to seek a caveat. The reason is telling. Even as early as the fall of 1874, Hubbard had learned that Elisha Gray, the well-known electrical researcher, had already filed a patent for a multiple-message telegraph system that overlapped with Bell's idea. Rather than tip their hand and lose priority with a caveat, Hubbard urged Bell to work as fast as possible and file a patent directly. That way, he and Pollok reasoned, they could try to undermine Gray's claim by arguing that Bell's conception for the invention had come prior to the filing of Gray's caveat.

By the fall of 1874, almost all the major players were in place and the stage set for the invention of the telephone. Bell quickly took Hubbard's advice to heart. He clearly recognized that to succeed he must produce his multiple telegraph before Gray did. As he put it on November 23, 1874:

> *It is a neck and neck race between Mr. Gray and myself who shall complete our apparatus first. He has the advantage over me in being a practical electrician—but I have reason to believe that I am better acquainted with the phenomena of sound than he is—so that I have an advantage there. . . . I feel I shall be seriously ill should I fail in this now I am so thoroughly wrought up.*

Of course, it is important to remember that the all-out race with Gray that Bell describes is *not* one for a telephone, but for the so-called multiple telegraph, capable of sending multiple messages simultaneously. Interestingly, though, it was around this time, in November of 1874, that the conception began taking shape in Bell's mind for a device to actually transmit speech. At first, Bell envisioned a receiver that was much like a sophisticated version of the talking machine he had built as a teenager with his brother Melly. Watson remembers that when Bell first spoke to him about the idea for a telegraph that could transmit speech, he described it as a machine, perhaps as big as an upright piano, that would simulate the transmitted vocal sounds using a multitude of tuned strings, reeds, and other vibrating mechanisms.

Not surprisingly, however, Gardiner Hubbard kept his gaze focused intently on the multiple telegraph, insisting that Bell focus on it exclusively. According to Watson's later account, no doubt colored to some extent by the benefit of hindsight, Hubbard told Bell to

> *perfect his telegraph, assuring him that then he would have money and time enough to play with his speech-by-telegraph vagary all he pleased. So we pegged away at the telegraph and dreamed about the other vastly more wonderful thing.*

WITH HUBBARD AND Pollok directing the operation, Bell would file his first patent related to his research on the multiple telegraph on March 6, 1875. Hubbard's strategic hand was everywhere to be seen in Bell's dealings with the U.S. Patent Office. But, as I researched Hubbard's role in Bell's patents, I learned that Hubbard's handling of Bell's later patent, for "Improvements in Telegraphy"—the one that would come to be known as the telephone patent—stood out in particular.

Some of the most detailed information about Bell's path to the telephone actually comes from legal testimony in patent lawsuits over the telephone's first decade. Bell, Hubbard, and others, fighting off legal challenges to Bell's patent, gave detailed testimony to substantiate their right to an exclusive claim over the telephone. Often their comments came in response to pointed questions about what had transpired. Bell's book-length deposition in one such case—now an exceedingly rare volume—was even published in 1908 by the Bell Telephone Company in an edition the firm described as the most complete account ever given of the telephone's development.

Unfortunately, historians have too often failed to mine these rich legal documents adequately. Bell's biographer Robert Bruce is something of an exception, but even he dismissively explains in the note accompanying his bibliography that

> *Most of the 149 volumes of printed testimony in litigation unsuccessfully challenging the Bell telephone patents deal with alleged*

> *inventions of the telephone independent of and prior to Bell and so has [*sic*] no bearing on his story.*

There is little question that these volumes of testimony make for often slow and painstaking reading, and I certainly cannot claim to have looked at most of their pages. But, based on the resources at my fingertips at the Dibner Institute's library, I soon found that the testimony of Bell, Hubbard, and others often had momentous bearing on the story of the telephone. In one lawsuit challenging the invention, for instance, Bell made a most extraordinary admission under oath about the timing of his patent: he testified that on February 14, 1876, Hubbard had filed the telephone patent himself, *against Bell's specific wishes and directions.*

Bell explains that he had explicitly directed Hubbard and his lawyers to wait while Bell sent an emissary and family friend—George Brown, an editor at the *Toronto Globe*—to file for a patent in Britain. At the time, the British Patent Office would issue a patent only if the specified technology had not already been patented in other countries.

The British patent was important to Bell: not only was he, at the time, a subject of the British crown, but such British rights were not covered by his agreement with Hubbard and Sanders. Thus, Bell worked out a separate agreement with the Brown brothers, George and Gordon, under which he stood to control a full half of the profit his technology might bring if it could be successfully commercialized in Great Britain.

George Brown had left by ship for Britain on January 25, 1876. On February 14, he had yet to cable about the matter. In Hubbard's testimony in one case, he claims that Bell:

> *did not hear from Mr. Brown as he expected, and finally wrote to me that if he did not hear by a certain day, that I might file it.*

In the voluminous archives of telephone arcana, no record survives of such a communication from Bell to Hubbard. More important, though, Bell's own testimony on the matter tells a notably different story. As Bell explains in his deposition:

> *Mr. Hubbard, becoming impatient at the delay, privately instructed my solicitors to file the specification in the American Patent Office, and on the fourteenth day of February, 1876, it was so filed without my knowledge or consent.*

Hubbard's involvement had intrigued me from the first. But this discovery, perhaps more than any other, confirmed my commitment—no matter what the effort—to unravel the true story of the telephone's development. Hubbard knew Bell had a formal, written agreement concerning his patent strategy in England. In fact, Hubbard was present at a meeting with Bell and George Brown during which they drew up a note to reflect their agreement. It read, in part:

> *It is understood that Mr. Bell will not perfect his applications in the American Patent Office until he hears from Mr. Brown, that he may do so without interfering with European patents.*

As a lawyer, Hubbard would surely have been unlikely to take it upon himself to break Bell's agreement lightly or frivolously. And yet, if Bell's testimony is to be believed, Hubbard never consulted Bell about the matter. Nor, if he was merely impatient, is there any evidence to indicate that Hubbard tried to take the simple step of sending a cable to Brown to inquire about the delay.

What prompted Hubbard to summarily negate Bell's prior agreement without his consent? The most likely explanation is that Hubbard felt it was urgent to file when he did. Especially given the Bell team's long-standing "neck and neck" competition with Gray, it strains credulity to imagine that Hubbard's hurried, unilateral action came only coincidentally on the exact day that Gray's caveat was filed.

As I pondered the information, it seemed far more likely that Hubbard was somehow tipped off about Gray's intention to file a caveat. The vague outlines of Hubbard and Pollok's clever telephone gambit were just starting to emerge, but many unanswered questions remained. If Hubbard was tipped off, how did he get the information? And why wouldn't Gray have objected once the Bell team's actions came to light? I lacked answers, but I knew one thing for sure: the timing of Bell's and Gray's telephone claims is normally portrayed as mere coincidence, but Hubbard's apparent rush to file—behind Bell's back, no less—strongly suggests otherwise.

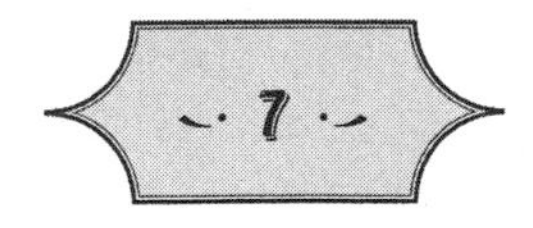

CLEAR RECEPTION

By late fall of 2004, I had amassed a growing number of questions about the invention of the telephone. I had listed them in my handwritten notebook, typed them into my laptop computer, and written them on notecards. My office came with a white board—the kind with erasable Magic Markers—and one day, on a whim, I used it to jot down some of my questions in the hope that, by looking at them together in one place, I could develop a more coherent research strategy:

Why didn't the U.S. Patent Office require Bell to submit a working model of his invention?
Why didn't the truth about Bell's apparent plagiarism come out in years of bitter court battles?
Why didn't Elisha Gray pursue his claim?

Before I knew it, I had nearly filled the entire board with questions like these when David Cahan, my colleague from the next office, knocked on my door.

"You busy?" he asked, glancing quizzically at the scribbling on my white board.

"No, no. Come in."

Cahan had come bearing a gift. He handed me a sheet of paper.

"I have no idea whether this is of interest," he began, "but I came across evidence that Bell and Helmholtz actually did meet face to face when Helmholtz visited the U.S. late in his life. I thought you might want to see it."

It was a photocopy of a newspaper article he had unearthed from the *New York Daily Tribune* dated October 4, 1893. According to the report, Bell had come immediately from Nova Scotia upon receiving word that Helmholtz, who had so inspired his work on the telephone, was in America. Ironically enough, the aged Helmholtz had decided to make the trip primarily to attend an International Electrical Congress organized and led by Elisha Gray. But, as the article reported, Bell did manage to meet his mentor in New York. The two men had lunch, after which Bell attended a lecture by Helmholtz and a reception held in his honor.

I was touched by Cahan's gesture of finding a link between his research and mine. I thanked him for thinking of me and he took a seat in the overstuffed chair across from my desk. He told me he was almost done writing a journal article about the impact of Helmholtz's visit to the United States. I told him I would be interested to read it.

"Listen," I said, in an impulsive overture that was doubtless long overdue. "I'm not ready to share this widely yet, but if you have a minute, there is something I'd love to get your advice about." Shuffling through the papers on my desk, I placed the photocopies of Gray's caveat and Bell's version of the liquid transmitter side by side on the corner of my desk and explained how I had happened upon them.

Cahan listened attentively and scrutinized the copies closely. Then he was silent for a long time.

"These are very intriguing documents," he said. "If the facts are just as you say, it would seem that you really could have something here."

He paused again. "Of course, there is more I would want to know. The key thing that comes to my mind is the danger of Whiggism. Do you know about Whiggism?"

Seeing the blank look on my face as I struggled to imagine what Tories and Whigs had to do with the invention of the telephone, Cahan proceeded in his soft-spoken and collegial way to offer me a learned thumbnail on historiography—the study of *the study of* history. "Whiggism," Cahan said, was the historical pitfall of not seeing things in their own context but rather judging the past by the norms or standards of the present. The term likely derived from the penchant of certain politically allegiant historians in Britain to write history in terms that favored their own party. In the history of science and technology, Cahan explained, "Whiggism" meant assuming knowledge that one's historical subjects would have lacked: giving undue credence to a theory, for instance, because you know it was ultimately proven true, or otherwise casting historical subjects as having acted for anachronistic reasons. As another colleague would later put it, "It's hard to avoid, but whenever possible you need to guard against reading history backwards."

As Cahan continued: "What could be going on here—I'm not saying it is likely but it is possible—is that Bell and Gray both depicted their inventions this way because at the time it was a standard way of doing so."

Unlikely or not, the thought had not crossed my mind.

"To avoid any threat of Whiggism creeping into your analysis, I'd recommend scouring through the textbooks of the day to make sure that a picture of a man's head leaning over like that wasn't some kind of a standard way of depicting any number of new inventions. At any rate, it's a base that I would want to cover on something like this."

Cahan, a specialist in this particular period, didn't hesitate before offering some highly specific recommendations.

"I'm pretty certain that Helmholtz's works never included diagrams like this," he said. "But I'd look particularly at the work of Lord Rayleigh, who was widely read in America. Also any primers on electricity and magnetism from the period.

"I'll certainly keep this to myself," he added, going to the door, "but please do keep me posted on what you come up with. You'll want to try to rule out any alternative explanation you can think of. But there's no question this is a very interesting find. Thanks for sharing it."

And with that, he returned to his office.

Before the week was out, I followed Cahan's suggestion. I flipped through Lord Rayleigh's classic *Theory of Sound* and many other period texts. There were no similar diagrams. I paged particularly carefully through Daniel Davis's *Manual of Magnetism* (1842) and J. Baile's *Wonders of Electricity* (1872). Bell and Watson mention both books as sources of inspiration and, delving into their texts, I could easily see why. Among other fascinating things, Baile's work actually predicts the invention of what it called an "acoustic telegraph," noting:

> *Some years hence, for all we know, we may be able to transmit the vocal message itself with the very inflection, tone and accent of the speaker.*

Still, for all the illustrations I found of electrical contraptions, people were hardly, if ever, depicted. Several of Davis's detailed illustrations include a disembodied hand resting upon a device, but none depict a person's head as Bell and Gray had done in their drawings.

I gradually became more confident that Bell and Gray had not appropriated some common form of diagram from the period. The more I scrutinized the two drawings, the more certain I became that they were primary documents that represented that rarity: a "smoking gun" that forces us to reevaluate our received understanding of a historical event. In this case, the drawings, now more than a century old, revealed a clear and discernible act of plagiarism—committed by Bell in his private laboratory notebook on the crucial eve of his success with the telephone.

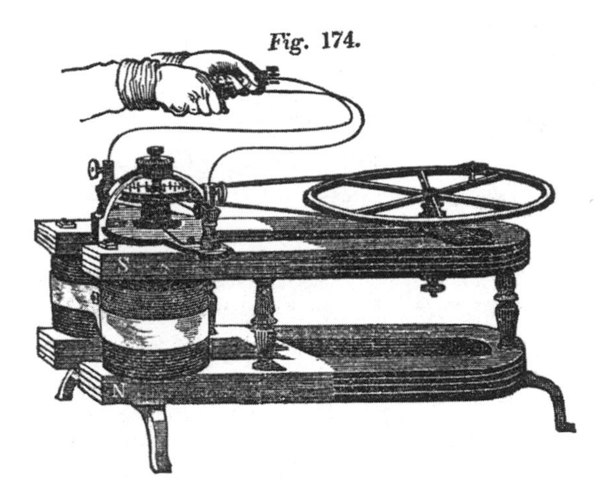

AN ILLUSTRATION FOR A SUPPOSEDLY THERAPEUTIC "ELECTROTHERAPY" DEVICE FROM DANIEL DAVIS'S *MANUAL OF MAGNETISM* (1847), A BOOK BELL MENTIONS AS A SOURCE OF INSPIRATION. NOTE THE DISEMBODIED HANDS AT TOP LEFT.

MORE THAN THE fear of Whiggism, what continued to nag at me was the widespread evidence of Bell's upstanding reputation. Those who knew Bell generally portray him as a gentleman of integrity with a keen sense of justice. In the preface to her biography, for instance, Bell's long-time assistant Catherine MacKenzie describes Bell as "honest, courageous" and "scornful of double-dealing." As MacKenzie writes,

> *The search for truth was the one really important thing in Bell's life. It is the irony of his story that the malicious charges of fraud, widespread against him during the long and determined effort to wrest the telephone from him, were in complete contradiction to everything essential in his character.*

MacKenzie, a close friend, assistant, and confidante of Bell's, could never be considered an unbiased source. Yet, while the passage made me all the more curious that charges of fraud against Bell were "widespread" in his day, my reading of Bell's notebooks and correspondence gave me no reason to doubt MacKenzie's fundamental assessment of his character.

I considered the possibility that Bell might have been unwittingly manipulated by Hubbard or his other advisers. But such a theory couldn't explain away the fact that the picture of Gray's transmitter had been drawn in Bell's own hand. At the very least, I reasoned, Bell must have cooperated with a plan to steal Gray's design even if he didn't instigate it. Given Bell's reputation for honesty, I couldn't help but wonder why.

Watson's autobiography offers a few potential clues. As with MacKenzie, there is no mistaking Watson's admiration for Bell. As he puts it:

> *No finer influence than Graham Bell ever came into my life. He was the first educated man I had ever known intimately and many of his ways delighted me.*

In their several years of close collaboration, the teacher opened up many horizons for his assistant. From the first, Bell encouraged Watson to learn algebra and introduced him to works by many of the leading scientists of the day, including Helmholtz. But, as Watson recalls, Bell's mentoring went far beyond the professional sphere. From Bell, Watson also learned how to comport himself like a gentleman, including everything from elocution to table manners.

Watson's abundant affection for Bell, like MacKenzie's, makes his autobiography rightly suspect in the eyes of most historians. So does the fact that it was written in 1926, toward the end of Watson's life, decades after the key facts it relates about the invention of the telephone. Despite these significant shortcomings, though, Watson's account is revealing for the detail he offers about Bell's unremitting frustration in his quest to build a multiple, or "harmonic" telegraph in

1874 and 1875. Try as they might, Watson and Bell could not make the device work. In Watson's words:

> *We accomplished little of practical value in spite of our hard work, the chief result attained being to prove to Bell that the harmonic telegraph was not as simple as it seemed.*

As Watson recalls, when the two rigged up the system in Bell's workshop in the attic above the Williams shop, the receiver would not respond to the signal reliably. Or it would respond when it wasn't supposed to. At best, the instruments were exceedingly difficult to tune to one another. According to Watson, things eventually got so bad that

> *my faith in the harmonic telegraph had vanished and, at last, after months of hard work on it, Bell's magnificent courage began to flag. I knew he was losing his enthusiasm for now I seldom heard his favorite expression, "Watson, we are on the verge of a great discovery," which, uttered, as it always had been, in a tone of conviction, would spur me on to renewed exertions to get improved apparatus finished and ready to try.*

Watson, in his account, emphasizes the strain on Bell from the arrangements he had made with his financial backers and the pressure Bell felt to succeed on their behalf. Sometime later, Bell himself wrote in one letter that he "never would have continued" in his research if not for his wish that Sanders and Hubbard "be repaid for the money they expended upon patents and upon my experiments."

In Watson's account, Bell's frustration and disappointment are palpable. But still I wondered: could the pressure have been great enough to lead Bell to steal a competitor's work? It certainly seemed at odds with Watson's glowing assessment of Bell's character and with most else I'd learned about him. As I soon came to appreciate, though, there was another aspect of the story to consider.

PERSON-TO-PERSON

IN THE WINTER OF 1874, Bell became a frequent visitor to the Hubbard home and a regular guest at the family's midday Sunday dinner. It was an elegant affair. Usually, the Hubbards served generous helpings of roast beef, followed by "floating island" for dessert—an almond-flavored meringue in a custard sauce that was particularly favored by Gertrude Hubbard. Bell no doubt relished the surroundings and the cuisine, but he welcomed the company even more. He soon found himself—awkwardly and privately—becoming increasingly infatuated with his winsome student Mabel Hubbard.

Bell's feelings for Mabel had begun to surface during their tutoring sessions. With Gardiner Hubbard often away for weeks or even months at a time in Washington, working on matters relating to the Hubbard Bill, and Gertrude Hubbard frequently accompanying her aging parents in New York, Mabel was often left with her older cousin, Mary Blatchford, in Cambridge. As an energetic and enterprising girl nearing her sixteenth birthday, Mabel would make her way from Cambridge to Boston by the horse-drawn streetcar for lessons with Bell or his teaching assistant Abby Locke in his office on Beacon Hill. Mabel recounts in her journal that when she first met Bell, she had found him interesting but "did not think him exactly a gentleman." His clothes were not fashionable, Mabel noted, and he dressed carelessly. At first, she had also guessed that Bell was over forty when he was, in fact, twenty-six years old. As the lessons progressed, though, Mabel's occasional letters to her mother show that she clearly warmed to her tutor.

Mabel wrote that Bell seemed to enjoy talking with her. He was full of so many ideas, Mabel said, that she rarely took her eyes off his face; she didn't want to miss any of them. As for Bell, Mabel charmed him from the first, and, on at least a few occasions, he found an excuse to walk with her to her ride after their class. Once, after a heavy snowfall, Mabel recounted, Bell

> *insisted on taking me to the streetcar. We had a grand time running downhill through deep snow. I was nearly up to my knees in snow but it was so dry I didn't get wet and the run kept me warmer than I generally am. I would have been almost sorry to get to the apothecary's but that I was quite out of breath; besides my waterproof and veil were flying about me and it was all I could do to hold on to them.*

In a postscript to another letter, Mabel writes:

> *What do you think, I have been told I am beautiful!*

And in another she tells her mother,

> *Mr. Bell said today my voice was naturally sweet. Think of that! If I can only learn to use it properly, perhaps I will yet rival you in sweetness of voice. He continues pleased with me. He said today that he could make me do anything he chose. I enjoy my lessons very much and am glad you want me to stay.*

With more opportunities to see Mabel outside of the classroom, Bell quickly came to view her less as his deaf pupil and more as the charming and appealing object of his affection. Shortly after Mabel turned sixteen, for instance, Bell attended a dance party held in her honor at the Hubbard's home. Despite the handicap of her deafness, Mabel hosted Bell and some twenty other young men and women with poise and grace, even though she was crestfallen to learn at the last minute that her dear friends Edith and Annie, daughters of Henry Wadsworth Longfellow, were both sick and couldn't come.

Gardiner Hubbard was away in Washington at the time, so Gertrude wrote him a full account of the party, describing what a strik-

Mabel Hubbard in 1870, at age thirteen.

ing young woman his daughter had become. As Gertrude recounted, Mabel received and introduced her guests

> *with the greatest ease and self possession . . . and one look at her face told how happy she was. I wish you could have seen her so fresh, so full of enjoyment and so very pretty. She wore her peach silk and looked her loveliest.*

With the first signs of spring in 1875, Bell was distracted and mystified by Mabel's growing hold on his affections. In a characteristic move, he started a special journal to help him sort out his feelings and chronicle their changing relationship. In it, he agonized over his situation and his emotions. Mabel was eleven years his junior. And Bell had been her teacher. He feared that she viewed him only in that capacity and would never accept him as a suitor. He fretted that a great gulf existed between his meager income and Mabel's affluent upbringing. He worried, too, that it would be improper for him to confide his feelings to one so young.

"I do not know how or why it is that Mabel has so won my heart," Bell wrote.

> *Had my* mind *chosen—or had others chosen for me—all would have been different. I should probably have sought one more mature than she is—one who could share with me those scientific pursuits that have always been my delight. However—my* heart *has chosen—and I cannot but think it is for the best—at least so far as I am concerned.*

If the cause of his affections was a mystery, Bell had no difficulty finding the quality in Mabel that he admired most. As he put it,

> *I value a gentle loving heart above all other things in this world—and I know that hers is such a one.*

ON JUNE 2, 1875, with the future of Bell's love life still very unresolved, he finally made a significant breakthrough in his telegraphic research after many months of frustration. Working in two adjoining rooms up in the attic of the Williams shop, Bell and Watson had set up a series of linked transmitters and receivers to test their latest version of the multiple telegraph. The circuit they built had three separate buzzerlike transmitters tuned to distinct pitches. Bell and Watson had also built six separate receivers, two for each transmitter; they placed three of them in Bell's room and three in the room next door. If things went according to plan, when Bell triggered a transmitter, the corresponding tuned receivers in each room—and only these corresponding receivers—would sound in sympathetic vibration.

After successfully tuning two of the transmitters to their matching receivers, Bell began work on the third set. But he could not get the corresponding receiver in the next room to function. Suspecting that its vibrating reed might have frozen against its contact, Bell removed the battery-driven transmitters from the circuit and called to Watson next door to pluck the transmitter's reed by hand to free it. Amazingly, even without the battery, Bell heard the reed in the receiver before him vibrate in concert with the one Watson had plucked. Bell yelled out to Watson next door to keep plucking all the reeds and, sure enough, Bell heard each of their distinct tones through the receivers in his room.

It was a momentous accident. Bell guessed correctly that the residual magnetism in the circuit had allowed the reeds to create just the kind of "tuned" electrical current he had been seeking for the multiple telegraph. If that were the case, it meant that his theory was largely correct, and that their repeated failures had resulted mainly from grossly overestimating the amount of vibration needed to send a message.

That day, as he listened to Watson pluck the reeds, Bell also realized something even more portentous. With his keen musical ear, Bell recognized that he could hear not just the pure tones created by the reeds' vibration, but also the overtones that gave them their particular timbre. As Bell would later contend, the discovery made him even more confident that it would be possible to send speech over a wire—if he could only figure out how to transmit it in the first place.

Bell was coming tantalizingly close to the invention of a telephone. Around this time, a letter to his parents clearly captures his sense of excitement and continued uncertainty:

> *I am like a man in a fog who is sure of his latitude and longitude. I know that I am close to the land for which I am bound and when the fog lifts I shall see it right before me.*

Unfortunately for Bell, the fog would linger for some time yet. Meanwhile, amid the important progress in his work, Mabel Hubbard occupied Bell's thoughts more than ever. And Watson began to notice the change in Bell, as he recounted much later in his autobiography:

> *I hadn't been in love since the time I was ten years old and had forgotten what an upsetting malady it could be until I observed its effect at this time on the professor. He was quite incapacitated for work much of the time. . . .*

At the end of June 1875, Bell was dismayed to learn that Mabel planned to vacation on Nantucket for the summer with her older cousin Mary Blatchford. The thought that Mabel would leave before Bell could confess his affection so troubled him that he resolved to take action. Alone in his Salem apartment, not knowing how else to handle the situation, Bell composed a letter to Mabel's mother. He wrote:

> *Pardon me for the liberty I take in addressing you at this time. I am in deep trouble, and can only go to you for advice.*

As Bell tried to delicately explain,

> *I have discovered that my interest in my dear pupil—Mabel—has ripened into a far deeper feeling. . . . I have learned to love her.*

Bell confided to Gertrude Hubbard that he wanted to tell Mabel of his feelings and learn whether she might reciprocate them. But he was well aware of Mabel's youth and, despite his strong feelings, he would do nothing against her parents' wishes. As Bell wrote,

> *Of course, I cannot tell what favour I may meet with in her eyes. But this I do know—that if devotion on my part can make her life any the happier—I am ready and willing to give my whole heart to her. . . . I am willing to be guided entirely by your advice, for I know that a mother's love will surely decide for the best interests of her child.*

Bell's letter prompted an immediate meeting with Gertrude Hubbard. Mrs. Hubbard liked Bell and undoubtedly tried to be kind and gentle. But she told him that she felt Mabel was too young to entertain thoughts of marriage. She believed her daughter needed time to mature, and she urged Bell to wait a year before telling Mabel herself of his feelings.

Several days later, upon returning from one of his many business trips to Washington, Gardiner Hubbard was even more adamant on the subject. As Bell summarized in his journal entry of June 27, 1875:

> *Called on Mr. Hubbard. Referred to my note of the 24th. Thought Mabel much too young. Did not want thoughts of love and marriage put into her head. If Mrs. Hubbard had not said one year, he would have said two.*

Having vowed to obey her parents' wishes, Bell now found himself in an emotionally excruciating predicament. He wanted more than ever to spend time with Mabel, but was explicitly constrained from betraying his true feelings to her.

Shortly before Mabel's planned departure to Nantucket, Bell's promise was put to an upsetting test. On a lovely June evening, he was strolling in the garden at the Hubbard's house with Mabel, her younger sister Berta, and an even younger cousin named Lina McCurdy. Berta and Lina were telling fortunes by pulling petals off a flower. When Bell's fortune came up "love," the girls teased him to confess who it was that had captured his heart. Flustered, but bound by his agreement to remain silent, Bell stiffly declined to answer. The dreadful result, Bell wrote later, was that he feared he had led Mabel to conclude he loved another and did not wish to tell her about it.

Once Mabel departed for her vacation, Bell's worry and regret about the incident grew to obsessive proportions. At the beginning of August, unable to bear it any longer, Bell went again to Mrs. Hubbard to announce his decision: he would go to Mabel in Nantucket and profess his love unless her parents explicitly forbade it. Mr. and Mrs. Hubbard tried to persuade him to wait at least until Mabel's return. But, just then, Mary Blatchford got wind of the news and took it upon herself to apprise Mabel of the situation.

On August 4, 1875, Mrs. Hubbard met once more with Bell. She explained that, just as he and her husband were hoping to discourage Bell again from acting precipitously, she had received a letter from Mabel which, she said, embarrassed her and left her "undecided how to act." By way of explanation, Gertrude Hubbard read out a portion of her daughter's letter to Bell. In it, Mabel asked pointedly whether Bell had requested her hand in marriage:

> *I think I am old enough to have a right to know if he spoke about it to you and papa. I know I am not much of a woman yet, but I feel very very much what this is to have as it were, my whole future life in my hands.*

Most impressive to Bell was the maturity evident in Mabel's letter. While she marveled at the prospect of Bell's affections and said she wasn't sure how she felt toward him, Mabel clearly relished the notion of making decisions for herself. As she wrote,

> *Oh it is such a grand thing to be a woman, a thinking, feeling and acting woman. But it is strange I don't feel at all as if I had won a man's love. Even if Mr. Bell does ask me, I shall not feel as if he did it through love.*

Mabel's apparent maturity bolstered Bell's resolve. He realized that his discussions with Mr. and Mrs. Hubbard, however well intentioned, had been unfair to Mabel herself. He pointed out to Mrs. Hubbard that

> *The letter which was read to me yesterday was not the production of a girl—but of a true noble-hearted woman—and she should be treated as such.*

As Bell put it, Mabel's wishes alone "should in future guide my actions." With that in mind, despite Gardiner Hubbard's warning that he would "regret this new burst of passion," Bell decided to go to Nantucket—with or without the Hubbards' blessing.

Traveling the whole day, Bell reached the Ocean House Inn in Nantucket by late afternoon. But a huge rainstorm made the island virtually impassable. So, in his hotel room, Bell penned a long, confessional letter to Mabel describing both his ardor and his promise to her parents to try to suppress his feelings until she was older.

> *You did not know, Mabel—you were utterly unconscious that I had long before learned to respect and to love you. . . . I have loved you with a passionate attachment that you cannot understand, and that is to myself new and incomprehensible.*

The next day, Cousin Mary informed Bell that Mabel would prefer not to see him but would be happy to accept his letter. Bell graciously acquiesced, glad to have at least written of his feelings so openly and at such length. After the tumultuous voyage, Bell headed home, still uncertain of Mabel's heart but feeling nonetheless unburdened.

Finally, at the end of the month, after Mabel's return to Cambridge on August 26, Bell got his long-sought chance to speak freely and privately with her. In the greenhouse behind her home, the two talked intimately and at length. Mabel told Bell she did not feel love for him or anyone—at least not what she described as the "hot" kind of love Bell had professed. But Mabel told Bell that she admired many things about him, did not dislike him, and welcomed the chance to get to know him better.

Bell could hardly have been more relieved and elated by the news. That very evening he closed out his journal by chronicling his feelings about Mabel, explaining:

> *Shall not record any more here. I feel that I have at last got to the end of all my troubles—and whatever happens I may now safely write: FINIS!*

BELL HAD DISPELLED his anxiety over his secret feelings for Mabel by bringing them into the open. But he had also, at this crucial juncture in his research, greatly complicated his business relationship with Gardiner Hubbard. There is little doubt of Bell's sincere love for Mabel. It is evidenced clearly in the way it literally drove Bell to distraction in the summer of 1875 and, even more, by Bell's later continued devotion to Mabel throughout the course of their eventual lifelong marriage. After the emotional summer of 1875, however, Bell had to worry not only about satisfying the hard-driving Hubbard in his telegraphic research, but also about staying in Hubbard's good graces in the hopes that he might one day become his son-in-law.

The task was not always easy.

In the fall of 1875, after an extended and recuperative visit with his family in Canada, Bell returned to Boston resolved to pay more attention to his ongoing work as a teacher of the deaf. His decision, no doubt influenced by his father, stemmed at least in part from his financial situation. Sanders and Hubbard had been supporting Bell's research and paying Watson's modest salary. But Bell had never asked his backers to help cover his living expenses, despite the fact that his research had cut deeply into the time he had available to earn a living.

Bell's decision greatly displeased Gardiner Hubbard. "I have been sorry to see how little interest you seem to take in telegraph matters," Hubbard wrote him that October, adding that Bell's behavior was "a very great disappointment" and "a sore trial."

Bell explained his need for income. And yet, given his newfound status as a suitor of Hubbard's daughter, he would not entertain any suggestion of a handout. As Bell wrote explicitly to Hubbard:

> *You are Mabel's father and I will not urge you to give—nor will I accept it if it is offered, any pecuniary assistance other than that we agreed upon before my affection for Mabel was known.*

Bell's pride aside, Hubbard was acutely aware that speed was essential if Bell was to beat out his competitors with his new telegraphic invention. So Hubbard used all the leverage at his disposal. He heatedly told Bell that, if he wanted to marry Mabel, he must give up his teaching and devote himself full time to research on the telegraph.

Well intended or not, Hubbard's ultimatum ignited Bell's temper and caused him to fire off an indignant response. As he wrote Hubbard,

> *I shall certainly not relinquish my profession until I find something more profitable (which will be difficult) nor until I have qualified others to work in the same field.*

Should Mabel come to love him in the way he loved her, she would surely accept him in any profession or business, provided it was "honorable and profitable."

Despite those strong words, Bell did redouble his efforts on his telegraphic research. Hubbard's bald threat may have been out of line, but Bell must surely have realized that a veiled version of it always remained: to further his relationship with Mabel, Bell would do best to carefully heed her father's business advice.

On November 25, 1875, Thanksgiving Day and the date of Mabel's eighteenth birthday, the two decided to become formally engaged. Bell, of course, was elated. But the decision no doubt raised some practical concerns for him as well. Bell certainly recognized the lavish lifestyle to which Mabel was accustomed and the huge gap between that lifestyle and his own means to provide it. He had euphemistically alluded to the problem in his initial confessional letter to Mrs. Hubbard, when he noted:

> *I know how young [Mabel] is and how many points of dissimilarity there are between us.*

To be sure, a paramount "point of dissimilarity" was the divergence between Bell's and Mabel's relative economic standing. That fact was exceedingly plain to Watson, who later wrote that

> *Professor Bell had a special trouble all to himself . . . he had fallen in love, wanted to get married and didn't have money enough.*

Mabel was "a charming girl," Watson added, and he could easily see why Bell fell in love with her. But

> *The important question with Bell was, where could he get the money that would enable him to marry?*

INTERFERENCE

So far, the strands of evidence all seemed to point in one direction. Bell was frustrated and discouraged in his effort to build a functional multiple telegraph system. He saw himself in an uncomfortably tight, high-stakes race with Elisha Gray. He was beholden, both financially and emotionally, to his hard-driving business partner and prospective father-in-law Gardiner Hubbard. And he was desperate to succeed financially to win wealthy young Mabel Hubbard as his bride. All these well-documented circumstances gave Bell ample motive to take unfair advantage of Gray, his competitor in the race to secure exclusive rights to the telephone.

Of course, establishing a motive is a far cry from proving complicity in a crime. The deeper I probed, the more I came to recognize that letters, contemporaneous accounts, and other primary sources offer powerful glimpses of past events, but paint a picture that is almost always incomplete. Motivations are often elusive, and the task of discerning them more than a century after the fact can feel almost impossible.

Bell's archive at the U.S. Library of Congress includes 147,000 documents—many of which are his voluminous personal correspondence. Yet despite the size of this trove of information, if Bell and his team *had* made a concerted effort to steal Gray's telephone design and falsely claim the credit for its invention, they probably would not have written explicitly to one another about the plan. And if by some chance they did, they would have been unlikely to retain the record of such correspondence. No, they would certainly have handled the matter as secretively as possible and destroyed all the evidence they could of any such actions.

How could I hope to overcome that obstacle?

"It is not an easy job," my friend and colleague Conevery Valencius counseled when I took up the issue with her, "but context is very important. You have to educate yourself enough to have the confidence to contextualize. For instance, in your case, it is valuable to know something about nineteenth-century letters, like the fact that, relative to today, there was tremendous circumspection and decorum in the way people expressed emotion."

Conevery's office stood just down the hall from mine. Trained at Stanford and Harvard, she is an extraordinary historian, whose book *The Health of the Country: How American Settlers Understood Themselves and Their Land* deservedly won an academic prize as the best environmental history of 2002. I had particularly sought out her help not just because of her credentials, but because of her disarming verve and lack of pretension. I had noticed at our group's regular seminars that she asked some of the most incisive and direct questions, while still managing to remain unflaggingly constructive and encouraging about her colleagues' research. In our small, rarefied group of researchers, Conevery, a native of Arkansas, often seemed to be the life of a very sedate party.

As I tried to make sense of the information I had found so far about Bell, I arranged to meet Conevery in her office to get her advice about how historians move from conjecture to proof in their interpretations of historical events. She was at work on a project about early Americans' understanding of geology; we sat amid large historical and topographical maps spread open on the floor. I told Conevery about my research and some of the pieces of evidence I had found. She modestly told me that she could suggest a number of colleagues who knew more about my material than she did, but I explained to her that it was not her factual knowledge I sought so much as her judgment as a researcher.

"Well, I guess what I'd say is that reading carefully is the main job of a historian," Conevery began. "And part of that means educating your own historical intuition so when you come across something unusual you can feel confident to say, 'This document or letter or journal entry seems different.' "

She offered an example of how intuition had paid off for her. In a close reading of the journals chronicling the explorations of Lewis and Clark, Conevery had spotted an intriguing passage recounting the fact that the legendary Indian guide Sacagawea—the only woman on the expedition—had become ill. Something about the passage seemed unusual to Conevery, and her historical intuition was especially piqued when she read Meriwether Lewis's entry noting that Sacagawea's illness occurred "in consequence of taking could." She had read enough letters from the eighteenth century to know that "taking a cold" was sometimes a euphemism for becoming pregnant. Conevery's hunch was that Sacagawea had a miscarriage. After finding supporting evidence, she wrote a paper along with a colleague offering this new interpretation. The theory helped explain many formerly mysterious things, such as why William Clark had inexplicably written that, if Sacagawea had died, it would have been the fault of her husband.

"Sometimes, when you consider an alternative interpretation for a historical event, a lot of disparate pieces that never held together well seem to fall more neatly into place," Conevery observed. "That said, though, as a historian you have to stick with and be true to your primary sources. They are your evidentiary core."

"In Bell's case," I told her, "I keep coming back to that incriminating sketch in his notebook. I could explain away many of my hunches about Bell's possible motivations, but I can't explain away that sketch." I pulled the copies of Bell's and Gray's drawings out of my briefcase. "What I know is this: the transmitter design depicted in Bell's drawing led him straight to a working telephone. I've tried, but I can only come up with one satisfactory explanation for its unmistakable likeness to Elisha Gray's proposed invention: and that is that Bell copied it from memory. If that's true, Bell must have seen Gray's confidential patent filing during his trip to Washington at the end of February 1876. I feel certain about it," I said. "But how can I hope to prove it?"

Conevery mused on the question for a moment and then peered at me, smiling.

"It seems like you're really asking two questions," she said. "First of all, my students often come to me and ask, 'Who am I to challenge the received wisdom about a historical event?' So I'll tell you what I tell them: 'That's your job. It's a big part of your job as a historian to interrogate your material and to trust your informed judgment about it.' It sounds to me like a part of your question is asking about your own authority here, and I would say you just have to believe in that and investigate this thing as honestly and thoroughly as you can.

"As for the more practical part, I might not be as much help. But it seems to me the key question you've laid out is a lot about patents. Bell's access to his competitor's material, from the sound of it, would likely have come either from his patent attorneys or someone at the Patent Office. If it were me, I'd probably start with the official documents and surviving supporting material about the patenting process itself."

I WAS GRATEFUL for Conevery's vote of confidence and her research advice. The patent angle was an obvious lead, and many of the documents to follow up on it were readily available. My natural starting point was to compare the actual patent claims filed by Bell and Gray on February 14, 1876. The distinction between them is stark indeed.

The caveat filed by Elisha Gray on February 14, 1876, is entitled "Instruments for Transmitting and Receiving Vocal Sounds Telegraphically." After the boilerplate introduction listing Gray's name and address, the caveat opens clearly and directly:

> *It is the object of my invention to transmit the tones of the human voice through a telegraphic circuit, and reproduce them at the receiving end of the line, so that actual conversations can be carried on by persons at a long distance apart.*

Now consider U.S. Patent No. 174,465—the famous telephone patent—filed by Alexander Graham Bell on the same day. Bell's patent is entitled "Improvements in Telegraphy." There is no question about its thrust: this patent describes Bell's efforts to create a telegraph system capable of sending multiple messages at the same time. As the patent states,

> *My present invention consists in the employment of a vibratory or undulatory current of electricity, in contradistinction to a merely intermittent or pulsatory current, and of a method of, and apparatus for, producing electrical undulations upon the line wire.*

After describing the benefits of this undulatory scheme in detail, Bell's patent explains:

> *Hence by these instruments two or more telegraphic signals or messages may be sent simultaneously over the same circuit without interfering with one another.*

In addition to this main purpose of the invention, however, Bell also notes:

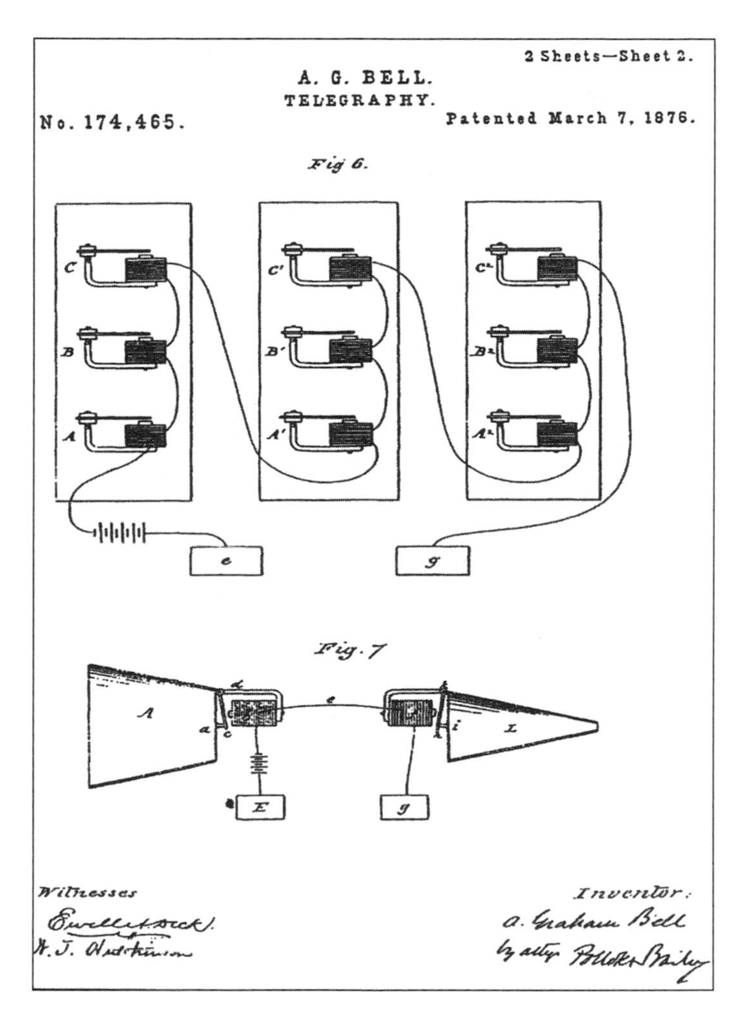

Diagrams from Bell's U.S. Patent 174,465, the so-called telephone patent, believed to be the most lucrative patent ever issued.

I desire here to remark that there are many other uses to which these instruments may be put, such as the simultaneous transmission of musical notes, differing in loudness as well as pitch,

> *and the* telegraphic transmission of noises or sounds of any kind *[emphasis added].*

Remarkably, this passage discussing "the telegraphic transmission of noises" is as close as the main body of the patent's text comes to describing a modern telephone. It makes no specific mention of transmitting speech—likely because, at the time of filing, Bell had no machine capable of doing so.

Meanwhile, though, patent lawyers often note that the heart of any patent filing lies in its claims. The two claims sections from Bell and Gray—normally listed at the end of the patent document—bear close comparison as well. Gray's filing contains a single claim. He states:

> *I claim as my invention the art of transmitting vocal sounds or conversations telegraphically through an electric circuit.*

Bell's patent, by contrast, has five claims. The first four claim the use of "undulations in a continuous voltaic circuit"—the invention Bell hoped would give him exclusive rights to a multiple-messaging telegraph. The fifth—and most famous—claim reads like an afterthought. Here, in the final sentence of the patent document, Bell makes his first and only mention of the transmission of vocal sounds by telegraph. The fifth claim asserts Bell's invention of

> *The method of, and apparatus for, transmitting vocal or other sounds telegraphically, as herein described, by causing electrical undulations, similar in form to the vibrations of the air accompanying the said vocal or other sounds, substantially as set forth.*

Bell's patent set my historical intuition on alert, especially given what I already knew. Even if Hubbard *had* encouraged Bell to focus on the multiple-messaging telegraph, something seemed strikingly awry. The most casual reader could see that one document describes an instrument to transmit speech, while the other describes the use of "an undulating current" to make a multiple-messaging telegraph work. I wondered why Bell's first explicit mention of anything akin to a telephone comes only in the final sentence of his claims section. Why wouldn't Bell have elaborated on the possibility of transmitting speech in the body of his patent? Was it possible that Bell tagged on the fifth claim *only after* he learned of the information contained in Gray's caveat? If that were the case, there may have been far more underhanded dealings than I had yet imagined.

THE OFFICIAL U.S. Patent Office filings themselves shed little light on the origin of the Bell patent's jarring fifth claim. However, I was able to find a good deal of information about the relative timing of Bell's and Gray's filings at the office. As I soon learned, controversy had raged over this matter practically from the start. With Bell's attorneys employing a highly questionable legal technicality, the timing turned out to be a key to Bell's legal victory over Elisha Gray. The issue of whose claim first reached the U.S. Patent Office figured prominently in court and even in a full-blown congressional investigation launched in 1886 to determine whether the U.S. government should annul the patent awarded to Bell. I was definitely not the first to suspect that there was something strange about the Patent Office's handling of these filings.

Mysteriously, to this day, the standard historical tale asserts that Gray's patent claim arrived at the Patent Office hours later than Bell's did. Almost all the major biographies of Bell offer a variant of this story. In 2006, for example, Charlotte Gray writes in her book *Reluctant Genius: Alexander Graham Bell and the Passion for Invention* that Gardiner Hubbard

> *dropped off at the U.S. Patent Office in Washington the application with its descriptions of Alec's [*sic*] electrical theories and of*

> *the crude apparatus that Alec and Watson had built. At first, Alec was furious with Hubbard for taking matters into his own hands. But Hubbard's action was crucial.* Two hours after Hubbard's call on the U.S. Patent Office, Elisha Gray filed a caveat there *[emphasis added].*

Despite the widespread repetition of versions of this tale, it is not grounded in fact. There is no evidence that Hubbard himself hand-delivered the patent that day; more important, there is no solid evidence to establish what time either document was filed on February 14, 1876. The simple reason for this is that the U.S. Patent Office officially noted only the date upon which documents arrived at its offices, not the time of day.

Much later, when Gray himself finally spoke out on the matter, he recalled having submitted his patent on the *morning* of February 14, 1876. As Gray explained:

> *The caveat was prepared deliberately, and completed the day before it was filed; and my recollection was that it was filed in the morning of Feb. 14. There was no reason for haste; I did not know or suspect that Bell was working on anything of the kind at that time. . . .*

A. Edward Evenson, an engineer in Illinois who has made a close study of this controversy, contends that, in fact, the evidence suggests Gray's caveat did arrive at the Patent Office first. Drawing upon the results of an investigation by the Department of the Interior in 1886, Evenson explains that, at the time, the procedure in the Patent Office was unvarying: all hand-delivered filings were placed in a large basket in the clerk's office over the course of the day. Each morning and afternoon, patent examiners would receive batches of patent applications in the mail. They would log them in a so-called cash blotter that served as the office's official log book, along with a record that the filing fee had been paid. After examiners had dispatched the applications that had arrived in the batches of mail, they would reach into the basket of hand-delivered applications and log these too, one by one, into the cash blotter. The fact that Gray's caveat is listed toward the end of that day's entries (the thirty-ninth entry) in the log book, while Bell's is listed as the fifth entry, means that Gray's filing was drawn from the bottom of the basket, *thereby likely having arrived early in the day.*

Piecing together the story from internal memos and the facts that surfaced in the ensuing legal case between Bell and Gray, Evenson presents his information in a fascinating book, *The Telephone Patent Conspiracy of 1876.* Evenson infers that the confusion over this issue stems from the fact that whoever delivered Bell's patent apparently insisted that it be immediately logged into the cash blotter and hand-delivered directly to the patent examiner in his office. We know this from correspondence between Bell's lawyers and Ellis Spear, who was the acting commissioner of the Patent Office at the time. The unusual hand delivery within the office accounts for the fact that, even though the documents were technically filed on the same day, the patent examiner in the case—a man named Zenas Wilber—received Bell's patent personally in his office on February 14 but did not receive Gray's caveat until the following day. But the available facts indicate strongly that Gray's claim arrived at the Patent Office several hours *before* Bell's, despite the oft-repeated story to the contrary.

The most important thing to note about the entire issue is this: *normally, it would make no difference whose claim was filed first.* At the time, the U.S. patent system was legally bound to issue patents to those who were the first to *invent,* not the first to *file.* Accordingly, after reading Bell's and Gray's filings, patent examiner Wilber properly ruled that, because of the overlapping claims of the two filings, Bell's application would be suspended for three months, allowing Gray time to formally file a full patent application, as the rules of the day stipulated. The Patent Office would wait until it had Gray's patent in hand before formally deciding whether to declare interference between the claims. Following this standard procedure, Wilber mailed notice of the temporary suspension (and likely interference) to Bell, Gray, and each of their attorneys on February 19, 1876.

This is the point in the story, however, where the contention over the time of day became vital to the telephone's future. Bell's attorneys, Anthony Pollok and Marcellus Bailey, immediately responded to Wilber's letter by arguing that if Bell's *patent* had been filed before Gray's *caveat* (which was not, after all, a full patent application), then no such suspension or declaration of interference should be allowed. It was a brash and highly questionable argument. As Pollok and Bailey no doubt knew, the Patent Office kept no official track of the time when applications arrived. In fact, just three weeks earlier, on February 3, in an almost identical case, Acting Patent Commissioner Ellis Spear had ruled against such an interpretation of U.S. patent law.

In that prior incident, known as the *Essex* case, Spear upheld the suspension of a patent application when an overlapping caveat was filed for a similar invention the same day. Like Wilber, the patent examiner in the *Essex* case—which dealt with an improved spinning machine invented by Jeremiah Essex—had made the standard office decision to call for a suspension in anticipation of formal interference hearings. As Spear wrote in that case:

> *There is nothing in the records of the Office to show which, if either, was, in point of fact, filed first.*

Therefore, Spear ruled that

> *I cannot take into consideration any representation of special hardship in this case, because it would be manifestly improper to consider any ex parte statement whatever. . . . I do not see why the law should not be strictly applied and the caveator notified and direct that it be done.*

Despite this clear and recent precedent, Pollok and Bailey made precisely the same plea in their own *ex parte,* or one-sided, case. In a letter to Spear received at the Patent Office on February 24, 1876, Pollok and Bailey wrote:

> *We respectfully request, before it is concluded to suspend our application for 3 months, that you determine whether or not our application was not filed prior to the caveat in question.*

Setting aside all other circumstances, the letter from Bell's attorneys seems suspicious to begin with. Pollok and Bailey would have been unlikely to go to the trouble of making this plea to Spear unless they felt confident that Bell's patent had, in fact, arrived first. Technically speaking, they should have had no way of knowing such a thing. And yet, as we know from the proceedings that followed, whoever delivered Bell's patent application had specifically bypassed normal procedure, insisting that it be hand-delivered to Zenas Wilber immediately. Could this have been part of a hastily made plan that could explain Hubbard's rush to file that day? It certainly seemed plausible.

In response to the letter from Pollok and Bailey, Acting Patent Commissioner Spear contacted Wilber about the matter. Wilber responded with the following memo:

> *The regular practice in the office has been to determine dates of filing by days alone, and in accordance with such practice I suspended the application herein referred to on a/c [account] of the caveat, the application and caveat being filed upon the same day, viz. Feby 14'—1876. In view of the practice above noted, I have paid no attention to the alleged differences between the times of the filings on the same day.*

Somehow, despite Wilber's explanation, and despite Spear's recent ruling in the virtually identical *Essex* case, Pollok and Bailey managed to prevail. They argued that Spear should disallow Gray's claim because Bell's patent application was logged into the cash blotter before Gray's caveat and because Wilber admitted receiving Bell's filing before Gray's.

In virtually all patent matters before and since, when two or more competing claims have arrived on the same day, the U.S. Patent Office has undertaken steps toward interference proceedings to determine which inventor conceived of the idea first. But on February 25, 1876, Acting Commissioner of Patents Ellis Spear made U.S. Patent Office history. At the written urging of Bell's lawyers, Spear overturned the initial judgment of the assigned patent examiner, ordering that only Bell's claim be considered because it had allegedly reached the Patent Office earlier in the day than Gray's caveat had.

Pollok and Bailey were clearly prominent and powerful players on the patent scene in this period, but it remains a mystery why Spear was swayed by their highly questionable argument. Even Spear's rationale is obscure. He specifically notes in his ruling that

> *Ordinarily the day of filing is not computed, and is considered* punctum temporis.

In Gray's case, however, Spear says it is within his right to consider "the exact time of day when an act was done," and, accordingly, he ruled that Gray's caveat "should be disregarded." Spear's reasoning may remain a mystery. But there is nothing unclear about its consequences for Elisha Gray: Spear's ruling drastically undermined his chances for legal recourse in the case.

The entire episode seemed extraordinary, even brazen in its irregularity, dovetailing with much else I had learned about the opening moves of the Bell team to assert their patent rights. I was relieved to find the work of Evenson and several others who shared my view that something was highly suspect in the patenting of the telephone. But I marveled that the controversy surrounding the telephone's origins was not more widely remembered today.

Along these lines, perhaps the best source of both information and outrage over the case occurred ten years after the fact, in an unusual governmental report. Dated December 22, 1885, the report was part of the congressional investigation into the anomalies in the case. At the time, enough controversy surrounded Bell's patent rights to the telephone that the government was considering annulling them. Because the U.S. Patent Office was then under the purview of the Department of the Interior, Assistant Secretary of the Interior George A. Jenks conducted an investigation of the matter for Congress. As it turned out, the investigation was politically tainted by the fact that some of the congressmembers who called for it stood to gain financially if the Bell telephone monopoly were dismantled. Nevertheless, the investigation makes for fascinating reading: it brings to light many of the particulars that the courts had previously ignored and stands as a thorough look at the entire patenting process in the case.

Jenks's report singles out Spear's decision in the Bell-Gray matter for particular attack, calling Spear's ruling "exceptional" and "contrary to the former practice of the office." Jenks notes too that Spear contradicted his own ruling on the same subject made on February 3, 1876, in the *Essex* case. Citing these and other disturbing irregularities, Jenks ends with this unequivocal assessment:

> *If in passing through a forest the woodsman should come upon the course of a tornado, and finds the tops of the trees all pointing in one direction he would be as firmly convinced of the direction the wind had blown as though he had been an eye witness to the storm. In this one-sided contest between the Bell application and the Gray caveat the tree tops all point one way.*

CALLER I.D.

ASIDE FROM THE wrangling at the Patent Office, I was curious about how Bell's work fit into the broader intellectual history of the telephone. For instance, even if he figured out how to make a working transmitter only after seeing Gray's caveat, I thought Bell might perhaps deserve credit as the first to have actively pursued the idea of a telephone.

I quickly realized, however, that such a notion was badly mistaken.

Efforts to trace the intellectual history of an idea or artifact almost always result in a complex tale with many disparate players. Surprisingly, though, unlike the patent question, most of those who have looked closely at the origins of the telephone agree about the key milestones that paved the way for its development. And almost all of them occurred well before Gray or Bell's time.

Over the course of more than a century, a number of people have carefully traced the telephone's conceptual development. George Prescott, a leading electrical researcher and contemporary of Bell's, did so in his 1878 book, *The Speaking Telegraph, Talking Phonograph and*

Other Novelties. The British electrical engineer William Aitken reexamined the history in 1939 in a detailed work entitled *Who Invented the Telephone?* And more recently, in 1995, Lewis Coe revisited the topic in *The Telephone and Its Several Inventors.*

Many of these experts contend, for instance, that much credit is due to Charles Grafton Page, an accomplished physicist and physician in Salem (the Massachusetts town where, coincidentally, Bell later lived with the Sanders family). In 1837, ten years before Bell was born, Page made a seminal breakthrough when he observed that an electromagnet emits a sound when the current flowing to it is interrupted rapidly. He called the effect "galvanic music" and found he could vary the sound by changing how fast the interruptions in current flow occurred. Most historians agree that Page was the first to use electricity to directly produce musical tones. His discovery opened the door to the prospect that electrical current could create sounds—and that the current could carry those sounds along a telegraph wire. This line of thinking would lead a diverse assortment of researchers on a direct path toward the telephone.

On numerous occasions, Bell referenced Page's discovery as a key influence on his work. He also frequently cited the importance of the electrical devices Helmholtz built to test his acoustical theories, especially the tuning fork sounder. In fact, in Bell's first public talk about the telephone at the American Academy of Arts and Sciences, he cites more than a dozen researchers who influenced his work in addition to Page and Helmholtz. As Bell himself notes, by the time he began his research, the acoustic effects related to Page's mysterious galvanic music had been

> *carefully studied by Marrian, Beatson, Gassiot, De la Rive, Matteucci, Guillemin, Wertheim, Wartmann, Janniar, Joule, Laborde, Legat, Reis, Poggendorf, Du Moncel, Delezenne, and others.*

The list may sound like an imposing historical tangle, but it offers some noteworthy context. Quite simply, by Bell's day many people, in many countries, were actively researching the intriguing, emerging nexus of acoustics and electricity.

Years later, in response to questioning in the lawsuits over patent rights to the telephone, Bell disavowed much knowledge of the work of any of the researchers listed above, at least so far as it might have helped him to invent a telephone; he claimed disingenuously that he didn't follow the literature in the field very closely. But Bell's later comments can to some extent be understood in the context of the adversarial tug and pull in the courtroom. The list Bell offered when he first publicly spoke about the telephone, on May 22, 1876, speaks for itself: Bell was at least to some degree familiar with the early telephonic research that preceded him.

With Bell's understandable focus on academic experimenters in his speech at the American Academy, though, he neglects to mention a particularly important conceptual breakthrough: an 1854 article by the Belgian-born engineer Charles Bourseul. Bourseul's article, "Transmission électrique de la parole" ("The Electric Transmission of Speech"), appeared in a French magazine called *L'Illustration*, offering what most historians believe to be the world's first published description of the telephone. "I have," Bourseul writes,

> *asked myself whether speech itself may be transmitted by electricity—in a word, if what is spoken in Vienna may not be heard in Paris.*

Bourseul didn't merely propose the notion of long-distance communication. He outlined the rudiments of how the system might actually work, describing an adaptation of the telegraph that he believed would allow it to transmit the human voice. As he explains:

> *Suppose that a man speaks near a movable disk, sufficiently flexible to lose none of the vibrations of the voice, that this disk alternately makes and breaks the currents from a battery: you may*

have at a distance another disk, which will simultaneously execute the same vibrations.

Aside from the faulty, telegraph-influenced notion that vocal sounds could be transmitted by "making and breaking" an electrical circuit, Bourseul presents a remarkably accurate description of the telephone's underlying principle: that sound waves on one end of a wire could be carried via an electric current and replicated on the other. He never attempted to build the system he described, but his article was widely read in scientific and engineering circles, and it inspired a number of researchers to pursue the idea. At the time of its publication, Alexander Graham Bell was seven years old.

"WELL, THERE'S SOMEONE up to no good."

I was crouched over my laptop computer at a Cambridge café when my colleague Dave Pantalony came by my table with a characteristic greeting. Pleased for the company, I asked him to join me.

Pantalony, an energetic and jocular postdoctoral fellow, was spending the year at the Dibner Institute after receiving his Ph.D. in history from the University of Toronto. Passionate about historic machines, he had already done a stint as a curator of a major U.S. collection of scientific instruments at Dartmouth College and he hoped for more such opportunities in the future. He asked what I was working on. I told him about my efforts to tease apart the intellectual history of the telephone.

"We've got to go together to inspect some of these early telephone devices," he said eagerly. Enthusiasm was a Pantalony hallmark.

"I'd love that."

I had spent enough time with folks at the Dibner to know that the field had long since fractured into two distinct camps: devotees of the history of science versus those who studied the history of technology. Although the distinction was hard for me to fathom, each group had its own associations, conventions, and journals. For his part, Pantalony was a strong proponent of the study of tools, instruments, and techniques. In a field dominated by philosophers tracing the development of theoretical knowledge, Pantalony took an almost archaeological approach. He believed that his colleagues would do well to pay more attention to scientific instruments and devices, rather than focus exclusively on documents.

In the course of his doctoral research, Pantalony had found an interesting way to further this view by focusing on Rudolph Koenig, one of the most skilled and meticulous instrument makers of the nineteenth century. Pantalony documented how Koenig, with a high-profile shop in Paris, made a discernible mark on numerous scientific fields by building instruments for scientific giants throughout Europe and North America, from Hermann von Helmholtz in Germany to Joseph Henry in the United States.

"The intellectual history of ideas can be like quicksand," Pantalony said. "But the machines—the instruments themselves—they tell their stories in such a concrete way. It's amazing how much you want to know is wrapped up in their design, materials, and construction."

We talked about a lot of things that afternoon, certainly more than we had in our many brief exchanges at the copy machine or office coffeemaker. And while we never did organize a joint visit to look at early telephone prototypes, I sought out Pantalony's expertise on several occasions as my research progressed. More important, though, I tried to take to heart his straightforward approach of emphasizing the tangible.

As a result of our chance encounter that day, I decided to take a trip to the Science Museum in London, which houses an unparalleled telecommunications collection that includes some of the world's earliest telephone prototypes. The pilgrimage seemed a far cry from my initial stroll around Boston in pursuit of physical traces of Bell's past. There was no escaping the fact that I had become deeply immersed into trying to unravel what I could about the telephone's inception.

When I arrived at the museum, John Liffen, curator of the communications collection, was my spry and knowledgeable guide. After meeting at his office in South Kensington, he whisked me away on an unforgettable journey to a vast warehouse on the outskirts of London. It was hard to keep up with both his whirlwind pace and his rapid-fire references to disparate topics in the history of technology. Liffen was encyclopedic, especially on details of the history of the telephone and telegraph.

When we reached the vast, unmarked brick building, once a postal facility, Liffen flashed his badge and chatted with the guard at the entrance. Then, as I scrambled along behind him, he swiftly climbed two flights of stairs to the rooms containing early telecommunications equipment. Through thick fire doors, we entered a vast warehouse filled with tall rows of heavy-duty metal shelves. They held the largest and most astonishing array of items I have ever seen.

We wove our way past many aisles. Some held aging architectural models of early factories and scientific instruments that looked like early vacuum chambers and static electricity generators. One aisle even featured a gangly and eerie assortment of historic prosthetic limbs made from wood, metal, and plastic. All told, it was a phenomenal graveyard of inventions, each a touchstone for little-known stories of human initiative and ingenuity. As a longtime casual collector of old tools and scientific instruments, I was enthralled by the surroundings. Despite regular visits, Liffen clearly sustained a similar passion.

"I never want to leave once I get here," he said.

We finally arrived at a cavernous wing of the warehouse devoted to communications devices. Our pace slowed a bit and I noticed that Liffen began to gaze at the collection as we walked past, as though fighting the temptation to explore and digress from our intended focus on the telephone. He was a specialist on the early telegraph, particularly the rare, very early "needle telegraphs" designed by the British team of William Cooke and Charles Wheatstone. The Science Museum has an unmatched collection of these remarkable devices, and, as we walked by, Liffen couldn't resist stopping briefly.

Sitting side by side on a waist-high shelf were a half dozen beautiful and intriguing needle telegraph machines. Encased in polished wood, with hand-carved detail, they looked like strange, Victorian-era clocks that might have once adorned somebody's mantelpiece. It was hard to imagine that, in their day, in the late 1830s, these radically new and avant-garde machines were the world's very first commercial electric telecommunication devices. Predating the simplicity of Morse code, the needle telegraphs made use of a fabulously idiosyncratic scheme in which telegraphic signals moved compass needles to spell out words by pointing to letters on a dial. Each machine had five compass needles with twenty letters arrayed around them in a diamond-shaped grid. The positions of the needles indicated a particular letter on the grid. As for the six letters of the alphabet missing from the grid, Liffen explained with a chuckle, the sender was out of luck; those letters simply had to be omitted from messages.

The needle telegraph machines were first employed for communication by trained technicians working at Britain's newly developing railways. By 1838 they were used to send telegrams between London and outlying towns. "On the one hand, given the advent of Morse code, these machines were obviously a dead end," Liffen mused. "And yet, on the other hand, what an extraordinary accomplishment they represent. They really can be seen to have paved the way for every telecommunications device that came after them."

Around the corner, Liffen paused briefly to pull down a length of now-fraying cable. It was a piece cut from the first transatlantic telegraph cable, laid in 1858, opening up the prospect of telegraphic communication between Europe and North America. The humble-looking cable, a little over an inch in diameter, was the physical representation of a truly enormous development in the history of human communication, shrinking the time it took to send a message from America to Europe from some two weeks by ship to a matter of mere minutes. We were working our way chronologically through the history of telecommunications, and the next aisle was devoted exclusively to the telephone.

I scanned the contents of the aisle as we approached, thinking immediately of Dave Pantalony's energetic passion for the artifact and the almost overpowering assemblage of clues contained here about the telephone's history. Floor to ceiling on either side of us were telephones and telephone-related paraphernalia. Some large pieces of switchboards caught my eye, as did a bright pink princess telephone from the 1950s. The museum's collection even included the "Osborne telephone" that Bell used in 1878 to demonstrate his invention to Queen Victoria (who was then staying at Osborne House on the Isle of Wight).

"I believe this is what you will be most interested in," Liffen said finally, pulling a compact wooden box off a shoulder-high shelf. He handed me a device designed by a German schoolteacher called Philipp Reis in 1863, when Bell was sixteen years old. "This," Liffen said, "is surely one of the very oldest telephones ever built."

I gingerly held the strange contraption to inspect it. Liffen was, of course, right. It was the item in the collection I had wanted most to see.

PHILIPP REIS'S STORY is intriguing. Other inventors may well have built early telephones that predated Bell's work, but none had a case as clear and well documented as Reis's. He taught physics at the Garnier Institute in the town of Friedrichsdorf, Germany, until his death at the age of forty, and during his lifetime, he remained largely outside the German scientific and technical elite. According to several accounts, Reis was most likely inspired to begin his research on a speaking telegraph device after reading Bourseul's article, either in its original 1854 version (Reis was fluent in French) or in a popular German translation published later that same year. Regardless of his inspiration, Reis also drew heavily, especially in his receiver design, upon the work of Charles Grafton Page. Reis had some rudimentary success sending sounds over telegraph wires as early as 1858, and dubbed his machine *das Telephon* from the Greek for "distance" (*tele*) and "sound" (*phon*), coining the term for the newfangled telecommunications device.

Over the next few years, Reis made a number of improvements to his device; by 1861, he had a telephone prototype that could reliably transmit music and at least some speech. He demonstrated it widely but never sought a patent on his invention.

Today, we know a good deal about Reis's story mostly because Silvanus Thompson, a respected English physics professor at the University of Bristol, wrote a detailed, book-length monograph celebrating Reis's work in 1883, some nine years after his death. It is unambiguously titled *Philipp Reis: Inventor of the Telephone.* Thompson examined Reis's telephone devices and writings with the analytical eye of a scientist and concluded that they were produced by the true inventor of the telephone. Notably, Thompson even tracked down witnesses who had seen and heard Reis's machine work during his lifetime.

One such witness was Heinrich Friedrich Peter, a music teacher at the Garnier Institute who was particularly interested in Reis's research. Peter visited Reis regularly in 1861, the year in which Reis first publicly demonstrated his telephone. Peter recounts that on October 26, 1861, he played the English horn while another colleague sang in a demonstration before a number of academics. He recalls that Reis's colleagues read sentences from a work by Adolf Spiess entitled the *Book of Gymnastics.* Reis, listening at the receiver, repeated the sentences to the audience as he heard them. When audience members protested that Reis must surely have known the sentences by heart, Herr Peter recalls that he personally

> *went up into the room where stood the telephone, and purposely uttered some nonsensical sentences, for instance: "Die Sonne is von Kupfer" (The sun is made of copper), which Reis understood as "Die Sonne ist von Zucker" (The sun is made of sugar) and "Das Pferd frisst keinen Gurkensalat" (The horse eats no cucumber-salad), which Reis understood [only] as "Das Pferd frisst."*

Peter's recollection is fascinating not only for its detail but also for what it shows of the limitations of Reis's device. For instance, his receiver, which drew directly from Page's research, was very weak and required the listener to place his or her ear up against the wooden box that held it. Even then, it was clearly hard to make out all the words. But Reis kept at it. One who witnessed his considerably improved model in action, according to Thompson's testimonials, was Georg Quincke, a renowned professor of physics at the University of Heidelberg. Quincke writes:

> *I was present at the Assembly of the German Naturalists' Association (Naturforscher Versammlung) held in the year 1864 in Giessen, when Mr. Philipp Reis, at that time teacher in the Garnier Institute at Friedrichsdorf, near Frankfort-on-the-Main, showed and explained to the assembly the telephone which he had invented. . . . I listened at the latter part of the apparatus, and heard distinctly both singing and talking. I distinctly remember having heard the words of the German poem, "Ach! du lieber Augustin, Alles ist hin!" etc.*

Not surprisingly, the demonstration "astonished and delighted" the association's members.

The 1863 model that Reis demonstrated to Quincke and others was the same one Liffen had just handed to me. The transmitter was a finely crafted, polished square wooden box. Two such boxes stacked atop one another would be about the size of a half-gallon milk carton. Protruding at an angle off one side of the wooden box was a metal speaking tube. Recessed into the top of the box was a thin, round metal diaphragm that sat just below a V-shaped metal bracket holding an electrical contact point at its vertex.

Liffen and I discussed the design. When the user spoke into the tube, it caused fluctuations in air pressure inside the box that would vibrate the metal diaphragm against the contact point on the bracket above. It was, in other words, a faithful realization of the machine Bourseul had envisioned a few years earlier. The reason it worked, however,

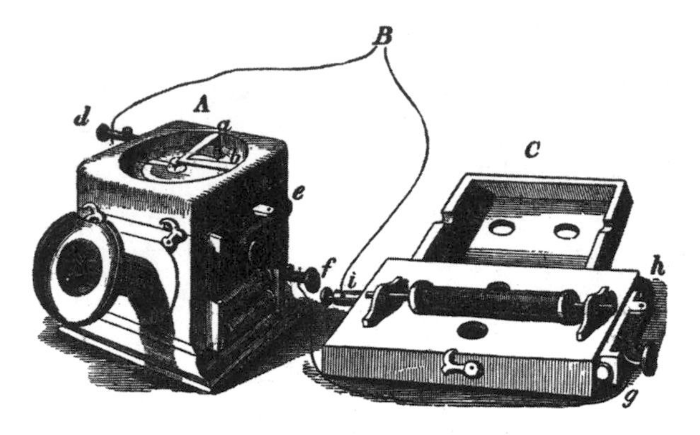

The telephone transmitter (*left*) and receiver (*right*) designed and built by Philipp Reis in Germany, 1863.

was notably distinct from anything Bourseul had envisioned. When the machine was adjusted just right, the diaphragm would not "make and break" contact with the bracket but would, if correctly adjusted, remain in loose contact with it, thereby allowing the sound waves of the speaker's voice to modulate the electric current from a battery and vary the resistance of the circuit.

"Most people don't believe that Reis fully understood what was going on in this device," Liffen said. "And yet, remarkably, it ended up working just about as well as Bell's model more than a decade later."

Reis demonstrated the 1863 model of his telephone fairly widely, and he sold at least seven machines to researchers in other countries. One, for instance, made its way to Stephen Yeates, a well-known instrument maker living in Dublin. Interestingly, Yeates not only successfully operated Reis's machine but significantly improved it, completely redesigning the receiver. In fact, Yeates's improved model, also part of the Science Museum's early telephone collection, sat on the shelf right next to the Reis telephone.

PANTALONY WAS RIGHT. The machines before me offered a refreshingly concrete starting point for any attempt to understand telephone history. I was holding in my hand a prototype reported to have transmitted music and speech while Bell was still a young teenager. Furthermore, it was distributed widely enough that knowledgeable practitioners even improved upon it. If that were the case, then why has Reis's contribution been virtually forgotten? More to the point: how much did Bell himself know about Reis's work? After all, Bell had been intensely interested in this rarefied field for years, and had contacted and corresponded with Alexander John Ellis, Sir Charles Wheatstone, and many others in Britain who traveled in scientific circles that would likely have been familiar with Reis's research.

Sure enough, with a little digging, I established that Bell did know quite a bit about Reis and his telephone. Bell's first public speech about the telephone at the American Academy of Arts and Sciences cites an article in which Reis discussed his telephonic research. Bell also cited a seminal article by Wilhelm von Legat, a Prussian engineer who described his experiments with Reis's telephone, explaining its design in detail. According to Silvanus Thompson, these were merely two of no fewer than fifty contemporaneous articles published about Reis's work. The chances are good that Bell read some of these in addition to the two he explicitly cited. For example, Bell could have learned all about Reis's device from the 1867 textbook *Electricity* by Robert Ferguson at the Edinburgh Institute in his own hometown. Ferguson doesn't just mention Reis in passing, he offers a picture of his telephone and a detailed two-page description of how the device operates.

There is also evidence that Bell spoke with others about Reis's work. In one important example, Charles Cross, a professor at the Massachusetts Institute of Technology and a colleague of Bell's, recalled under oath that he believed he had spoken to Bell about Reis's telephone on two occasions, including one as early as the spring of 1874, nearly two years before Bell's telephone patent.

Even more intriguing is the evidence that Bell not only read or heard about Reis's machine but that he saw it, perhaps as early as 1862, when he was fifteen years old. An 1886 article carefully tracing Reis's work notes that one of his early model telephones was purchased and demonstrated that December by a dealer in a well-known Edinburgh shop handling scientific equipment. Although Bell had gone to live for several months with his grandfather in London around that time, Bell's family was still based in Edinburgh. No concrete evidence survives about the matter in the voluminous Bell family correspondence, but one would imagine that Bell and his acoustically minded family would have been keenly interested in a well-publicized demonstration and sale of Reis's dramatic, novel machine in their own city.

Meanwhile, no imagining is needed to appreciate the fact that Bell received a firsthand demonstration of a Reis telephone in March 1875, almost a full year before filing his telephone patent. The incident occurred when Bell visited the laboratory of Joseph Henry at the Smithsonian Institution in Washington, D.C. According to the standard Bell biographies, Henry, then one of the world's eminent electrical researchers and a person who had played a major role in the invention of the telegraph, offered encouraging words when Bell paid him a call to demonstrate his early research on the multiple telegraph. In court, Bell admitted to having seen the device while visiting Henry's office at the Smithsonian Institution, although he tried to downplay its significance. As he noted,

> *Before March 7, 1876, I saw the complete Reis apparatus which Professor Henry had at the Smithsonian Institution. I had also read some publications referring to Reis, but I cannot say now what they were.*

All these connections, substantial and tenuous, leave little doubt that Bell was fully aware of Reis's device, an invention that was famously finicky, but was widely known for years to have successfully transmitted music and speech. As Reis himself wrote in 1863 to William Ladd, a well-known instrument maker in London, in a set of instructions accompanying his *Telephon*:

> *Any sound will be reproduced, if strong enough to set the membrane in motion.*

The extent to which knowledge of Reis's work guided Bell's thinking is unclear. There is no hard evidence, for instance, that Bell experimented directly with Reis's telephone, as he did with many of the devices, such as the tuning fork sounder designed by Helmholtz. Still, at the very least, Bell's knowledge of Reis's work ought to have tempered his claims to being the first with a working telephone. But it did not. As Bell declared in one of the many court proceedings:

> *I take the ground that all the experiments of Reis and others, and all the knowledge of persons skilled in the arts of acoustics and electricity, together with all the information contained in printed publications prior to March 7, 1876, as a matter of fact, failed to enable anyone to transmit intelligible speech with any apparatus at all, operated by electricity, until I showed how it could be done in my patent of March 7, 1876, No. 174,465.*

Amazingly, Bell seems to have prevailed with this self-serving view. On a key occasion in an early legal case when the issue of Reis's priority came up and a demonstration in a U.S. courtroom was actually staged, no one present managed to get Reis's instrument to successfully transmit speech.

I wondered what to make of that failed courtroom demonstration, whether it lent credence to the possibility that Thompson and the many testimonials he collected overstated the case for Reis. How well *did* Reis's machine work? There is no doubt that his device was finicky. But if the prevailing assessment of several experts is correct, Reis made a telephone in the 1860s that worked in spite of its incorrect conceptual reliance on the "make or break" conception spelled out by Bourseul. In communication via the telegraph, after all, the electrical circuit between the two ends of the communication is turned on and off and on in a pattern to spell out a message; hence, it is often called a "make or break" circuit. But such a connection cannot effectively transmit the sound of a voice. Rather, a working telephone (at least in Bell's day, a century before digitization) required a constant connection in which sound waves from a human voice would, via some kind of transmitter, vary the resistance in a continuous electrical circuit.

In his book, Silvanus Thompson argues that Reis must surely have possessed a working understanding of the principle we now call "variable resistance" in order to have built his machine. Thompson contends that Reis's telephone itself displays such an understanding, even if Reis's writings lack the vocabulary to effectively describe it.

I wasn't sure about Thompson's assessment. Perhaps Bell, with his patent's elaborate descriptions of "undulating current," rightly deserved credit as the first to comprehend this principle. But I knew that Dave Pantalony would probably argue that all the theorizing in the world would be of little use compared with the obvious option of testing Reis's device itself to determine how well it worked. After all, how much would Reis's accomplishment be diminished if he had made a working telephone without a full command of the underlying principle on which it operated? It is commonplace for pathbreaking inventions—especially, say, in the pharmaceutical field—to precede a thorough scientific understanding of what makes them work.

Despite the Reis telephone's all-important failure in U.S. court, I thought I had read in Aitken's book that the British Post Office in the 1930s had tested models of the Reis telephone and determined that they could, in fact, transmit intelligible speech. I asked Liffen about the matter over coffee in the employees' lounge at the Science Museum's warehouse, and the question seemed to make him uncomfortable. Eventually, he told me why.

In September 2003, he said, he was looking through a file cabinet in the museum's archives when he stumbled upon a document discussing the results of tests on the Reis telephone that had never been made public. In 1947, on the occasion of the centenary of Bell's birth, the Science Museum in London had worked with the British firm Standard Telephones & Cables (STC) to conduct a detailed series of experiments on the museum's Reis telephone. The company's engineers judged Reis's cigar-box receiver too weak to aid in their assessment, so they tested Reis's transmitter with a modern, loudspeaker-type receiver. The STC engineers found that Reis's finicky old transmitter worked perfectly. Next, simply amplifying the receiver, they found that it too received articulate speech clearly. Despite their initial intent to laud Bell on his 100th birthday, the report's authors concluded, based on their tests, that Bell could not accurately be considered the first to have invented a telephone capable of transmitting speech.

At the time, however, as Liffen explained, STC was negotiating a business deal with AT&T, the direct descendant of the original Bell Telephone Company. STC executives were evidently so afraid the study's conclusions might upset their corporate deal that they shelved the report and prevailed upon the museum not to discuss the matter. Liffen's predecessor at the Science Museum complained, but, presumably fearing to alienate a major corporate sponsor, he ultimately acquiesced in helping to hide the results. Despite the study's historical and educational value, the museum made no mention of the experiment or the report. Once again, it seems, history conspired to deny Reis's work the credit it deserved.

"It was not the museum's finest hour, I'm afraid," Liffen noted sheepishly.

TAPPING THE PHONE

THE LIBRARY AT the Dibner Institute generously allowed me to borrow most books in its collection for as long as I wished during the academic year. Some books, in its "vault" collection, however, were either so rare, old, or fragile that the library allowed access to them only in the reading room. Silvanus Thompson's *Philipp Reis: Inventor of the Telephone*, published in London in 1883, was one such book. In my many hours poring over it, I thought a lot about Thompson's vital role in spreading word of Reis's achievement. His rare, aging volume seemed like a remarkably thin and fragile thread connecting Reis's day to our own. Few copies of the book remain. Without Thompson, it is unclear how much, if any, information about Reis would have survived to the present.

Yet even with Thompson's detailed and meticulous biography, it is easy to see how Reis's circumstances limited the public attention and acclaim the humble schoolteacher received in his life and afterward. Reis was modest and relatively poor, and connected neither to those with political power nor to those with scientific expertise. And he died before his fortieth birthday.

Still, what about Elisha Gray? If my hunch was right, Gray, with his liquid transmitter, was the first to knowingly incorporate the concept of variable resistance into his design, a vitally important development in the history of the telephone. Given such a contribution, I found it hard to understand how history might have come to slight Gray.

He was, after all, one of the nation's premier electrical engineers, well respected and well connected to the powerbrokers of his day. He kept a close eye on developments in the field of telegraphy, and had invented everything from an improved burglar alarm to a "telautograph," a device we now know as a fax machine. Furthermore, by 1874, Gray was affluent enough to devote all his time to independent research and invention and to protect his work with the help of legal counsel. And he lived for many years after his pathbreaking work on the telephone—certainly long enough to have tended to his legacy.

Gray's lack of recognition was a mystery I found difficult to unravel, at least partly because a dearth of information has survived about him. I did, however, find some telling glimpses of Gray's life and times. One came in a forty-three-page booklet—another of the Dibner's "vault" holdings—commemorating a banquet held to honor Gray in 1878. Thrown by friends and admirers in his hometown, the affluent Chicago suburb of Highland Park, it was a lavish affair. Hundreds of guests attended, an orchestra played, and a host of flowery and long-winded speeches regaled the audience after the elegant sit-down dinner. Most notable was the banquet's stated purpose: to laud Gray *for his invention of the telephone.* In his own day, at least, Gray seemed to have won some recognition.

One Chicago newspaper editorialized:

> *The citizens of Highland Park gave a banquet on last Friday evening in honor of Dr. Elisha Gray, the inventor of the telephone.*

> *. . . Dr. Gray has met the fate which has so often overtaken great discoverers, in the attempt made to deprive him of both the honor and profits of the achievement; but which, we are happy to say, will not be successful in this case.*

The evening's toasts hit many similar notes. For instance, Gray's friend, S. R. Bingham, a prominent Highland Park lawyer who helped organize the event, told the assembled guests:

> *If the press and the public have been misled—either by the willingness of other men to wear borrowed laurels, or the reluctance of our modest friend to demand his own—it is high time that we give to the press and the public authenticated facts.*

Among the facts Bingham offered was a firsthand account of having learned of Gray's "musical telephone" (which could transmit music but not yet intelligible speech) as early as the summer of 1874. In December of that year, Bingham recalled, he attended a public exhibition of Gray's musical telephone at the Presbyterian church in Highland Park; astonished parishioners became the first sizable audience in America to listen to music electronically "piped in" from another room. At the time, Bell had barely begun his research on the telephone.

The booklet about Gray's banquet is also filled with excerpts from congratulatory telegrams that had arrived from all quarters to mark the occasion. Apparently, local friends like Bingham were not the only ones to credit Gray with the invention. In just one notable example, Western Union's chief electrician, George Prescott, addressed his message to: "Elisha Gray, the inventor of the telephone and solver of the problem of the ages."

ONE OF THE pleasures I found to break the routine of working in the Dibner's tranquil reading room was the chance for an occasional chat with the library's research director, David McGee. A historian of science and technology with an impressive breadth of knowledge, McGee was irrepressibly cheerful and self-effacing. Hailing from Canada, he was also a self-described "Bell admirer." Bell spent much of his time in his later years at his estate in Canada and is almost universally revered there. McGee liked to gently tease me that my research was likely to make me unpopular "back home" if I stressed the accomplishments of inventors other than Bell.

Joking aside, McGee knew so much about Bell that he could eagerly and effectively rebut any but my most carefully researched doubts about Bell's accomplishments. At one point, he even went to the trouble, in order to challenge some of the theories I had begun to develop, of typing up a detailed listing of Bell's and Gray's major, known actions between 1875 and 1877. It was a lucid and helpful document that must have taken him hours to compile; but, in characteristic fashion, McGee wrapped his gift with humor, playfully calling his document the "Bell Crime Labs Timeline."

McGee was poking light fun, of course, but his gibe also rang true. I was increasingly convinced that a crime *had* taken place—a blatant and immensely consequential one. And the actions of the victim—Elisha Gray—were particularly hard to understand. When Bell laid claim to an invention so nearly identical to Gray's prior conception, why didn't Gray cry foul?

Bell admirers like McGee have emphasized, rightly, that Gray was too focused on improvements to the telegraph to appreciate the commercial potential of the telephone early on. To be sure, in 1875 and 1876, Gray's attention, like Gardiner Hubbard's, was directed intently upon the goal of commercializing the multiple-messaging telegraph. Perhaps because of this, Bell biographer Robert Bruce goes so far as to argue that

> *in coming up with his telephone idea, Gray broke away more sharply from his current line of thought and research than had Bell. This fact adds probability to what was certainly possible as*

early as September 1875: that some general hints, if not details, of Bell's new goal had reached Gray and vibrated in his mind like the sympathetic response of a tuned reed.

THERE IS PLENTY of evidence that Bell and Gray kept tabs on each other's research. But I was highly skeptical about Bruce's inference, however poetically phrased, that Bell had led Gray to his research on the telephone. From what I had gleaned, the opposite was at least as likely to be true.

According to Gray's own account, his path to the telephone began early in 1874, when he found his nephew playing with some electrical equipment in the bathroom. At the time, many households had so-called vibrating rheotomes hooked up to the bulky batteries of the day, designed to administer electric shocks for treating everything from muscle pain to asthma. As was later shown, the treatments had no medical value, but in the 1870s "electrotherapy" machines were popular. Using such a setup, Gray's nephew was "taking shocks" to entertain a small group of younger children (whether they were siblings, cousins, or friends is unclear from Gray's account). For the demonstration, Gray's nephew had hooked a wire from the machine's induction coil to the zinc bathtub. He held the other lead in his hand, and then completed the circuit by rubbing his free hand against the tub.

When Gray walked in on the show, he was fascinated: the vibrating induction coil emitted a sound when the boy touched the tub. Experimenting further along a line of inquiry once pioneered by Page, Gray found that he could change the tone emanating from the machine (what Page had dubbed "galvanic music") by adjusting the rheotome's frequency. Furthermore, he could make the sound louder by rubbing the metal tub harder and more quickly. Over the next few months, Gray experimented in earnest with the effect. Soon, he had built a strikingly modern version of a telephone receiver that could play musical tones. Gray used his discovery to transmit these sounds

Elisha Gray's experiments with his bathtub in 1874 led to his invention of a so-called musical telephone.

via a telegraph wire. In the spring of 1874, Gray showed the device to Western Union officials and to Joseph Henry and other scientists at the Smithsonian Institution.

Word of Gray's musical telephone was widely reported in newspapers that summer. As a practical inventor and entrepreneur who specialized in improvements to the telegraph, Gray naturally began to think of applications for his device. He quickly experimented to demonstrate that his receiver could pick up audible tones produced by signals carried for hundreds of miles over a telegraph wire. Like Bell, he realized he might be able to incorporate the phenomenon into a telegraph capable of carrying multiple messages simultaneously. And, much to Bell's consternation, as demonstrated in his comment about the "neck and neck" race between the two, Gray patented his version of a harmonic, multiple telegraph slightly before Bell did in 1875.

The idea of transmitting vocal sounds was a natural extension of Gray's research. As I had learned from reading about Reis and others, the notion of a "speaking telephone" had been discussed for years in the scientific literature.

Gray said that the specific impetus for his moving in that direction came in 1875, when he saw two boys in Milwaukee playing with a tin-can-and-thread telephone. When the boy at one end spoke into the can, the sound waves from his voice would mechanically travel along the string to the boy listening at the other end. As soon as he noticed it, Gray said, it dawned on him that the sound vibrations might be carried electrically. In his caveat of February 1876, leaping beyond the research of Reis and others, Gray became the first to describe a groundbreaking way to accomplish this.

Gray's idea for a liquid transmitter drew directly upon the concept of a "water rheostat" that his Western Electric Company had produced several years earlier. In this earlier invention, the resistance in an electric circuit could be decreased or increased by lowering or raising a platinum strip in a liquid solution.

Still, the opinions of many historians like Bruce—highlighting Bell's better-known work and belittling Gray's contributions—convinced me that I needed to learn more about Gray for myself to determine the full extent of his claim to the invention of the modern telephone. Unlike Bell's papers, though, which are superbly cataloged at the U.S. Library of Congress, documents pertaining to Gray's work are far fewer in number and scattered around the country. Furthermore, in contrast to the countless biographies of Bell, little published work exists about Gray's life and work.

Piecing together available biographical information, and tracing the notes and references in several secondary texts on the telephone, I learned that Oberlin College held a number of Gray's papers. I assumed at first that Gray must have left them to his alma mater, but then I stumbled across a different explanation. In an intriguing passage, Lewis Coe, the author of the 1995 work, *The Telephone and Its Several Inventors*, notes that

> *One of Gray's staunchest supporters came forth in 1937 in the person of Dr. Lloyd W. Taylor, head of the physics department of Oberlin College. Dr. Taylor was convinced that Gray was the real inventor of the telephone, even though Bell held the legal claim.*

According to Coe, Taylor, a careful researcher, had personally tracked down many of Gray's original documents and brought them to Oberlin. Coe even reprints Taylor's one published article about Gray as an appendix to his book. Entitled "The Untold Story of the Telephone," it appeared in the December 1937 issue of the *American Physics Teacher.* As soon as I read it, I felt a rush of excitement familiar to any historical researcher hot on a promising trail.

Taylor authoritatively addressed many nagging questions about Elisha Gray. First of all, Taylor validated my hunch about the importance of Gray's liquid transmitter. As Taylor put it, Gray's 1876 caveat

> *was the first embodiment of the principle of variable resistance applied to the telephone, and as such possesses an historical importance which can scarcely be overemphasized.*

Taylor closely evaluated the timing of Bell's and Gray's respective technical accomplishments, and I was impressed by his diligent scholarship. Among other things, he explains clearly that the transmitter Bell used to call Watson on March 10, 1876, was very different from the instrument described and illustrated in Bell's patent. Furthermore,

Taylor documents that the type of liquid transmitter Bell first used had previously been described by Gray

> *in a confidential document about the contents of which Bell subsequently acknowledged having received information.*

Bell acknowledged having learned of Gray's caveat? This assertion certainly piqued my interest. Even more tantalizing, according to Lewis Coe, Taylor had been at work on a book-length manuscript about Gray when he died in a mountain-climbing accident on Mount St. Helens in July 1948. As far as I could tell, Taylor's unpublished manuscript and its source materials had collected dust at the Oberlin College Archives ever since. I couldn't help but wonder whether Taylor's research materials might offer the same kind of vital link to Gray that Silvanus Thompson's had previously provided for Reis's research. I decided on the spot to make the trip to Oberlin to examine Dr. Taylor's unpublished manuscript and source materials in person.

HISTORIC AND FIERCELY independent Oberlin College dominates the sleepy Ohio town that shares its name, a half hour southwest of Cleveland. The college's archive is housed atop a squat, fortresslike library at the heart of the campus, accessible only by a special, keyed elevator behind the reference desk. I was excited to reach the place, but it seemed like an awfully remote and unlikely venue to hold a potential key to the story of the telephone's invention.

Curator Roland Baumann, a soft-spoken, avuncular historian who oversees the collection, was warm and welcoming. He's been on the job for almost twenty years. After that amount of time, he said, "when you walk in those stacks, the documents talk back to you."

Despite the archive's quiet and sequestered air, Baumann says his office fields some 1,700 to 1,800 research requests per year. Many of them pertain, in one way or another, to Oberlin's extraordinary history as the first college in the nation to accept women (since its founding in 1833) and African Americans (since 1835). For many years prior to the Civil War, in fact, Oberlin was the only institution of higher learning in the nation a black woman could attend. Not surprisingly, given that history, Oberlin continued, until well after the Civil War, to educate more blacks than any other college in America.

Baumann and I talked about the papers I had come to see. Baumann said he didn't know the particular documents well, but he did know that many of the most important materials about Gray were filed with Taylor's papers. This, he said, reflected how Gray's documents had come into the archive's collection. The issue of "*provenance,*" which Baumann pronounced with a studied French accent, "is vitally important to historians, and especially to archivists," he said. By way of explanation, he pulled a notebook off a nearby shelf to show me the archive's accession data. He could tell me, he said proudly, when each group of documents had come into the collection and under what circumstances.

"We need to keep records on the chain of custody of these resources because historians are always looking for 'smoking guns.' They need to know where that document or letter comes from to help judge its authenticity. As I try to tell the administration here, 'we are all about evidence.' But I'm not sure they really understand that."

Ken Grossi, Baumann's assistant, produced three boxes of papers and two compact discs of digitized holdings in the sizable Taylor collection to get me started. I had come a long way to read Taylor's manuscript. Wading through his papers to get to it, I learned a good deal about Taylor himself.

As the former head of the physics department, Lloyd W. Taylor taught at Oberlin College for more than two decades, from 1924 until his death in 1948. Over the course of many years, he had become seriously preoccupied with Elisha Gray. Reading through his papers, it occurred to me that Taylor personally identified with Gray. Both had been modest, straitlaced midwesterners. Gray held a long-standing position at the physics department of the Oberlin faculty, giving lectures there occasionally. Gray had been a religious man and a teetotaler. So was Taylor, whose wife was prominent in Ohio's temperance movement during the era of Prohibition. Whatever the cause of his interest, though, Taylor's meticulous historical research and his technical knowledge as a scientist give his analysis a good deal of credibility.

Taylor was convinced that Bell had plagiarized Gray's liquid transmitter design. He arrived at this conclusion after noting that Bell hardly mentioned the possibility of a liquid transmitter in his 1876 patent. By Taylor's generous count, Bell's oblique references to it makes up just eight out of a total of about 190 printed lines. Furthermore, Taylor writes,

> *There is no suggestion anywhere that up to this time Bell had ever made, or contemplated making a liquid transmitter.*

Taylor's analysis goes considerably beyond the controversy over the liquid transmitter, however. In closely investigating Gray's and Bell's receiver designs, Taylor concludes definitively that Gray

> *made and publicly used several types of telephone receiver many months before Bell constructed his first one.*

Many historians contend that Bell had priority in his invention of the telephone, including the telephone receiver. Taylor's careful analysis effectively refutes this view. With primary evidence, including his own tests on some of Gray's original devices themselves, Taylor makes a convincing case that Gray's priority is perhaps even more pronounced in his receiver designs than it is with the liquid transmitter. As Taylor explains, Bell was concerned with trying to make receivers that would respond to only one frequency in his multiplex telegraph; as a result, it was not until the late spring and summer of 1875 that Bell first attempted to construct anything that could be considered a modern, electromagnetic telephone receiver.

Gray, meanwhile, had constructed and publicly demonstrated relatively sophisticated receiver designs for his musical telephone in 1874 and early 1875, a year earlier than Bell. As Taylor explains,

> *Gray had made and exhibited to some hundreds of witnesses who were qualified to comprehend their principle and importance, four types of telephone receiver, all of which possess thin metal diaphragms and hence anticipated the design of the modern telephone receiver much more closely than did the receiver which Bell first devised in 1875.*

At this time, Gray, like Bell, had yet to figure out how to transmit intelligible speech. Importantly, though, Taylor's analysis shows that three out of four of Gray's receiver designs were fully capable of receiving speech. As a result, he concludes, Gray had clear priority over Bell in his designs for both a receiver and transmitter. Speaking of the modern telephone receiver and the vital liquid transmitter, Taylor writes:

> *Gray's loss of credit for these two major contributions to the development of the telephone was quite possibly due in part to his own ineptitude as a tactician. As to the facts of his priority in both fields there is little room for controversy.*

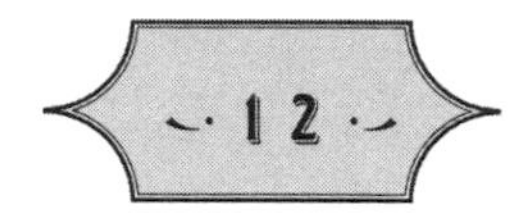

BAD CONNECTION

TAYLOR'S MANUSCRIPT, MOSTLY complete but never published, was a scholarly and illuminating read, but his source material was little short of a gold mine. Conducting his research in earnest in the 1930s, Taylor contacted Gray's descendants and gained access to many formerly unavailable documents. He seemed to have thrown himself into the project over the course of many years. He corresponded with a variety of people who had known Gray or otherwise been connected to the inventor, and he tirelessly wrote to encyclopedias and other publications in an attempt to resuscitate Gray's reputation. In response to his efforts, the editor of the *Encyclopaedia Britannica* even agreed to commission Taylor to write a new entry on Gray for inclusion in their section on telephone history.

Perhaps most fascinating, Taylor retrieved a long-forgotten trove of Gray's papers from the attic of a house where Gray had once worked. As Taylor's collected correspondence reveals, when he first got word of these newly unearthed documents, he appealed to Gray's relatives to allow him to bring the papers to the Oberlin Archives.

Among the gems of the collection is a revealing letter to Gray from Alexander Graham Bell, complete with a hand-addressed envelope, dated March 2, 1877. As I pieced together later, correspondence between the two had begun several days earlier, when Gray requested permission to demonstrate Bell's telephone design, along with his own, at a public forum, while offering to give Bell full credit for its design. While the initial query from Gray testifies to his gentlemanly decorum, Bell's hasty reply by telegram denies Gray permission unless he is willing to repudiate comments that had been printed in the *Chicago Tribune* questioning Bell's priority as the telephone's inventor.

In this follow-up letter at the Oberlin Archives, however, Bell apologizes for the earlier telegram in which he had lashed out at Gray. By way of explanation for his temper, Bell claims, with at least some exaggeration, that his priority on the telephone results from the fact that his patent documents for the telephone had been completed "for months" as he waited to file in England. He also explicitly says that he knew nothing about Gray's work to construct a telephone for the purpose of transmitting vocal sounds. Elaborating on this point, Bell makes a most revealing disclosure. He writes:

> *I do not know the nature of the application for a caveat to which you have referred . . . except that it had something to do with the vibration of a wire in water—and therefore conflicted with my patent.*

As of March 1877, Gray's filing was still not a public document; it had never been published. In this letter, Bell admits to Gray, presumably inadvertently, that he had knowledge of Gray's work—and how could that knowledge have been anything but illicit? What's more, Bell knew the most important detail about Gray's liquid transmitter design, namely, the use of a needle in water to convert sound waves to changes in electrical resistance. Bell's admission offers solid corroboration to go with the sketch in his notebook that he had intimate knowledge of the details of Gray's confidential caveat.

How did he get it?

There in the boxes of the Taylor Papers, I found an astonishing answer in a formal-looking document, penned in blue ink on roughly a dozen bound, lined sheets of thick paper, notarized and bearing multiple signatures. The document had evidently come into Gray's possession at some point and thus into Taylor's hands. I was thankful for Baumann's care with regard to issues of provenance, because this document was a smoking gun if ever there was one. It was an affidavit dated April 8, 1886, signed in Washington, D.C., by Zenas Fisk Wilber, the patent examiner in charge of telegraphy. Wilber had personally handled Bell's telephone patent and Gray's caveat.

In a clear, neat hand the document states:

> *Zenas Fisk Wilber being duly sworn deposes and says:*
>
> *I am the same Zenas Fisk Wilber who was the principal examiner in the United States Patent Office in charge of a division embracing all applications for patents relating to electrical inventions, during the years 1875, 1876 and till May 1st 1877 about which latter date I was promoted to be Examiner of interferences; that as such examiner in the applications of Alexander Graham Bell, upon which was granted to him Letters Patent of the United States No. 174,465, dated March 7th, 1876, for "Multiple Telegraphy," was referred to me and was by me personally examined and passed to issue.*

Wilber also acknowledges that he has previously made sworn statements about his role in the matters he is about to discuss. In none of his previous statements, he says, had he previously told the entire truth about the circumstances connected with the issuance of Bell's telephone patent.

> *In order that justice may be vindicated and injustice rectified, I have concluded to tell the whole truth and nothing but the truth. . . . I am fully aware that it may place me in an awkward position with some of my friends and possibly before the public, that it may even alienate some of my friends from me, nevertheless I have concluded to do as above stated, regardless of consequences without the hope or promise of reward or favor on the one hand, and without fear of results on the other hand. This affidavit is consequently the outcome of a changed mode of life and a desire on my part to aid in righting a great wrong done to an innocent man.*

There, in a sequestered archive in Ohio, I was more than intrigued by the document before me. I was transported. The document continued:

> *I am convinced by my action while Examiner of Patents that Elisha Gray was deprived of proper opportunity to establish his right to the invention of the telephone and I now propose to tell how it was done.*

Wilber explains that he was an alcoholic and that he owed money to Marcellus Bailey, the second partner in the law firm that filed Bell's patent. He says that he had known Bailey for some thirteen years. They served in the same regiment and as staff officers of the same brigade commander in the Union Army during the Civil War. After the war, both Wilber and Bailey came to Washington. Wilber began work at the U.S. Patent Office and Bailey enrolled in the Columbian College Law Department (now George Washington University Law School), graduating in its first class, in 1866, and going on to a successful patent law practice.

In addition to their close personal acquaintance, Wilber notes, he had borrowed money from Bailey on several occasions, even though the commissioner of the Patent Office had explicitly prohibited such arrangements between patent examiners and the attorneys with whom they worked.

Wilber says that when he first saw the collision between Bell's patent and Gray's caveat, he followed the appropriate rules by initially suspending Bell's application. After sending the suspension letter on February 19, 1876, however, Wilber says that Bailey visited him to ask about the situation. Feeling beholden, Wilber says that, in a flagrant violation of Patent Office rules, he told Bailey the facts about Gray's caveat, thereby helping him to immediately craft a protest against the suspension on Bell's behalf. Then, also out of a feeling of indebtedness to Bailey, Wilber says, he neglected to undertake a thorough investigation, as requested by the acting patent commissioner, Ellis Spear, to determine which document had been filed first at the Patent Office. Instead, Wilber says, he simply settled on the notion that Bell's patent had arrived first.

Wilber then states that, in an act of even more historic consequence, Bailey pressured him to let Bell see Gray's confidential caveat during Bell's visit to Washington on February 26, 1876. As he explains,

> *Professor Bell called upon me in person at the office, and I showed him the original drawing of Gray's caveat, and I fully explained Gray's method of transmitting and receiving. Prof. Bell was with me an hour, when I showed him the drawing and explained Gray's methods to him.*

In the affidavit, Wilber even adds a diagram depicting the layout of the office showing where everything in his version of events had transpired. Wilber claims that Bell returned to the office at two o'clock in the afternoon that same day and, in the hallway outside his examiners' room, presented him with a hundred-dollar bill. Wilber notes:

> *I am fully aware that this statement will be denied by Prof. Bell and that probably the statements I have made as to my relations with Maj. Bailey and his influence will be denied, but nevertheless they are true, and they are stated, subscribed and sworn to by me*

1

City of Washington } s.s.
District of Columbia }

Zenas Fisk Wilber being duly sworn deposes and says;

I am the same Zenas Fisk Wilber who was the principal examiner in the United States Patent Office in charge of a Division embracing all applications for patents relating to electrical inventions, during the years 1875, 1876 and till May 1st 1877, about which latter date I was promoted to be Examiner of Interferences; that as such examiner the application of Alexander Graham Bell, upon which was granted to him Letters Patent of the United States No. 174,465 dated March 7th 1876, for "Multiple Telegraphy," was referred to me and was by me personally examined and passed to issue.

And I am the same Zenas Fisk Wilber who has given affidavits in the telephone controversy, commonly called the "Bell controversy," in which a bill has been filed and suit brought in the Southern District of Ohio by the United States for the voidance of Letters Patents No. 174,465 issued March 7th 1876 and No 186,787 issued January 30th 1877, both to Alexander Graham Bell, which affidavits so given by me were used at the hearing before the Commission in the Interior Department, consisting of Secretary Lamar, Assistant Secretaries Muldrow and Jenks and Commissioner of Patents Montgomery, which commission sat for the purpose of advising the Department of Justice as to the advisability and propriety of the General Government bringing the suit noted supra.

The first page of the final affidavit of former patent examiner Zenas Fisk Wilber, sworn on April 8, 1886, and uncovered by Lloyd Taylor in 1933.

while my mind is clear and my conscience is active and bent on rectifying as far as possible any wrong I may have done.

I SOON DISCOVERED that Wilber's affidavit is extremely problematic both as a legal and a historical document. First of all, Wilber's sworn statements, made in conjunction with the congressional investigation into irregularities in Bell's patenting of the telephone, occurred long after most of the legal cases challenging Bell's patent had worked their way through the courts. Even if Wilber had come forth earlier, though, it wouldn't have changed the fact that the patent examiner had made a number of contradictory sworn statements on the matter. In fact, between July 30, 1885, and the final one signed on April 8, 1886, Wilber executed no fewer than five separate affidavits, and several of them directly contradict one another. That fact alone, coupled with Wilber's admitted alcoholism, would have made his confession all but worthless in court.

In typically meticulous fashion, however, Lloyd Taylor analyzed all of Wilber's extant statements and even offered several of them in an appendix. He shows that Wilber's affidavits are not as contradictory as they might first appear. As Taylor explains, most of Wilber's statements prior to his final affidavit merely leave out parts of a story he ultimately divulges. For example, all except one of Wilber's statements at least mildly charge that Bell and his lawyers engaged in fraudulent activity. Wilber's sworn statement of October 10, 1885, for instance, explains how Bell's lawyers manipulated the normal delivery process to make it appear as though Bell's application had reached the Patent Office before Gray's did.

One statement by Wilber, made on October 21, 1885, does take a notably different tack in defending himself against charges of wrongdoing, a fact that would certainly seem to undermine his credibility. Nevertheless, months later, in his final affidavit on April 8, 1886, Wilber explains the circumstances. On October 21, he says, a lawyer for the Bell Telephone Company had sought a statement from him. At the time, Wilber contends,

> *my faculties were not in their normal condition and I was in effect duped to sign it, not fully realizing then, as I do now, the statements therein contained. I had been drinking, was mentally depressed, nervous and not in a fit condition for so important a matter.*

To be sure, Wilber, an alcoholic and an admittedly corrupt government official, is not the upstanding witness one might hope for to provide a definitive account of such a matter. His track record in making incomplete and contradictory statements raises even more questions about his reliability as a source. Nonetheless, it is hard, after reviewing Wilber's statements, not to feel swayed by the heartfelt and confessional quality of his final affidavit. Zenas Wilber stood as the pivotal insider with the ability to facilitate Bell's plagiarism. He certainly had nothing to gain from divulging such an incriminating story. More important, though, Wilber's detailed final accounting of events, offered in April 1886, dovetails so closely with a veritable trove of other circumstantial evidence that it cannot be easily discounted. It offers, for instance, the first plausible explanation I had found for how a drawing of Gray's liquid transmitter wound up in Bell's notebook just days after his return from Washington, D.C.

As for the provenance of the document, Taylor notes that he found it himself in 1933 in the attic of Elisha Gray's old laboratory in Highland Park, Illinois. When he discovered it, Taylor explains in his notes, the original of Wilber's last affidavit had a slip of paper clipped to it with the following comment written in pencil in handwriting he believed to be Elisha Gray's:

> *Extract from a letter dated Aug. 24, 1889 from a friend of Wilber's (when boys) just after Wilber's death.*

Dedicated sleuth that he was, Taylor even located the letter Gray's note referred to. It had been stored separately in the attic of Gray's house in Highland Park, and was first found by Frederick W. Cushing, a friend of the Gray family. The letter, dated August 24, 1889, had been sent to Gray by a Major Marion Van Horn after Wilber's death, presumably along with the affidavit itself. The final lines of Wilber's affidavit explain the situation:

> *I have thus concluded after full frank consultation and conversation with my old college mate, comrade in arms and long time friend, Major Marion D. Van Horn and I shall entrust this document to him, hoping and trusting that I may yet be able to repair in some degree the wrongs done and I stand ready and shall always be ready and willing to verify this statement before any court or proper tribunal in the land.*

In the letter, addressed to Gray care of his company address, Van Horn stated that he felt Gray should have the original of the document. His letter ends:

> *In conclusion I will state that Mr. Wilber, whether drunk or sober, always maintained that his [final] affidavit is a true story!*

ON THE LINE

WILBER'S AFFIDAVIT, PUBLISHED in *The Washington Post* on May 22, 1886, prompted a swift and adamant public denial. Three days later, on May 25, the newspaper printed Bell's sworn retort:

> *So far as my personal acts or knowledge are concerned I know that all proceedings in the filing and prosecution of the application for my part and in the grant of it were free from fraud and trickery and honest in all respects; and I believe that all the acts of others concerned were also in all respects honest.*

Bell categorically denied most of Wilber's specific allegations. He said he had never given Wilber a hundred-dollar bill or any payment. And he added flatly:

> *Mr. Wilber did not show me Gray's caveat or the drawings of it or any portion of either.*

Wilber's affidavits, coming nearly a decade after the fact, never played a role in any of the legal challenges to Bell's claim to the telephone. But Wilber's final affidavit, in particular, bears a close reading alongside those numerous other occasions when Bell spoke specifically about the matter.

On several occasions, Bell simply denied having had any knowledge whatsoever of Gray's caveat. One such comment, for instance, came late in Bell's life, when his son-in-law Gilbert Grosvenor sought to write his authorized biography. Grosvenor embarked on the project in 1905 and pursued it for over a decade. During that time, he also served as the editor of *National Geographic* magazine and, given this and many other duties, he never managed to complete the book. Nevertheless, the Library of Congress contains one fascinating document, dating most likely from around 1910, in which Bell and his wife Mabel each offer extensive typed comments in response to an early, fragmentary draft from their son-in-law. Referring to Grosvenor's handling of the controversy surrounding Gray's caveat, Bell writes:

> *I knew that some interference with my patent had been declared through some misunderstanding—but this interference had been withdrawn before I reached Washington, and I could not find anything out about it, as a caveat was a confidential document. I shrewdly suspected Gray, for we were working on parallel lines in developing musical telegraphy at the time that the transmission of the human voice was concerned. . . . While I feared Gray might have struck vocal sounds in his caveat*—I knew nothing of the contents of the documents until Gray himself referred to the matter in his letter asking permission to exhibit my apparatus *[emphasis added]*.

Of course, having previously reviewed the 1877 correspondence between Bell and Gray, I knew that Bell must have had some knowledge of Gray's caveat because he wrote that he believed Gray's claim had something to do with the "vibration of a wire in water." Furthermore, as always, the picture of Gray's device in Bell's notebook nagged at me. After all, it had been drawn almost immediately upon Bell's arrival home in Boston after having met with Wilber, Pollok, and Bailey in Washington. For me, Bell's sketch hung like a cloud over all the various iterations of his story, casting many of his claims into deep shadow.

Continuing to look closely into the matter, I found ever more inconsistencies. Perhaps the most troubling occurs in one of the first telephone patent cases, the *Dowd* case of 1879, which ultimately led to a historic settlement between the fledgling Bell Telephone Company and the established telegraph giant Western Union. In response to questions, Bell's testimony offers a strikingly different version of events—one significantly closer to the story Wilber recounts in his final affidavit. Speaking under oath, Bell admits that he *did* discuss Gray's caveat with Wilber. By way of explanation, Bell contends that, during his visit to the patent office in February 1876, Wilber had pointed to a particular paragraph *in Bell's own patent application* to explain the overlap between his and Gray's claims. According to Bell,

> *As I knew nothing of the matter, I asked the Examiner what the point of interference had been. He [Wilber] told me that a caveat was a confidential document, and therefore declined to show it to me, and I did not see the caveat nor any part of it, but the Examiner indicated in a general way the point of interference* by pointing to a paragraph in my application *of February 14, 1876, with which I understood him to say the caveat had conflicted [emphasis added].*

In particular, Bell says,

> *It was a paragraph in the body of the specification, and read as follows: "Undulations are caused in a continuous voltaic current by the vibration or motion of bodies capable of inductive action; or*

by the vibration of the conducting wire itself in the neighborhood of such bodies. . . . The external resistance may also be varied. For instance, let mercury or some other liquid form part of a voltaic circuit, the more deeply the conducting wire is immersed in the mercury or other liquid, the less resistance does the liquid offer to the passage of the current."

The paragraph Bell mentions in his testimony—the one he says Wilber pointed to—does seem to anticipate the principle, now known as variable resistance, that makes Gray's transmitter work. With its reference to a wire immersed in a liquid, it also appears to anticipate the very method used in Gray's liquid transmitter. Like so many other aspects of the telephone patent saga, it seems suspiciously coincidental that Bell would have independently invented the very mechanism that made Gray's telephone work, especially because virtually nothing in Bell's laboratory notebook suggests that he had experimented with the notion prior to filing his patent application.

Nevertheless, taking Bell at his word that Wilber had only pointed out a paragraph in Bell's own patent, the passage above still stands as a most remarkable admission. In it, Bell acknowledges under oath that he received specific, confidential information about Gray's work before he had ever successfully transmitted intelligible speech via the telephone.

Given the benefit of hindsight, Bell's account of the hint Wilber gave him appears even more questionable in light of information that emerged in the *Dowd* case: in Bell's original file copy of his patent application, the particular paragraph in question regarding the concept of so-called variable resistance—the one Bell claims Wilber pointed to—*is written into the margin of the document*, presumably added sometime after the rest of the application had already been written.

Once again, because truth *is*, as they say, stranger than fiction, I offer below a picture to persuade the skeptical. Again, this version of the patent application, from the Library of Congress, is Bell's original file copy. The version of Bell's patent that went to the U.S. Patent Office was copied over by a professional copyist in Pollok & Bailey's office before being formally submitted. In this original version, however, a paragraph is written into the left-hand margin seemingly in Bell's handwriting. The passage contains Bell's reference to variable resistance—the seminal claim that made the telephone possible and that ultimately swayed the courts to uphold Bell's patent against the many legal challenges it faced in its first decade. This is also the paragraph Bell claims Wilber pointed out to explain where his application had conflicted with Gray's caveat.

Call it historical intuition, or just a journalist's horse sense, but, to me, Bell's testimony simply contains too many overlapping coincidences to be convincing. Given the central importance of variable resistance, is it plausible to believe that Bell left such a crucial passage out and then independently added it into the margin as an afterthought? As the British engineer John Kingsbury put it as early as 1915:

> *Strange, isn't it, that an inventor should omit till the last moment . . . the essential feature of his invention?*

Two decades later, William Aitken echoed the sentiment, likening Bell's omission of the variable resistance clause to writing a play "without the leading character."

So, when did Bell pen the crucial addition into his patent application? The question goes to the heart of the central mystery: namely, how much did Bell independently know about variable resistance and how much did he actually learn from Gray's caveat?

The question leads into one of the deepest thickets yet.

BELL SPENT THE evening of January 12, 1876, at the Hubbard's house in Cambridge. That evening, Mabel, now Bell's fiancée, was feeling ill. Her mother and sister were away for a social engagement and

neighborhood of another wire – an undulatory
current of electricity is induced in the latter.
When a cylinder upon which are arranged
bar-magnets.. is made to rotate in front of the
pole of an electro-magnet an undulatory
current of electricity is induced in the ~~latter~~
coils of the electro-magnet
Undulations ~~may be~~ are caused in a continuous
voltaic current by the vibration or motion of bodies
capable of inductive action; – or by the vibration
of the conducting wire itself in the neighborhood
of such bodies. X

In illustration of the method of
creating electrical ~~currents~~ undulations, I shall
show and describe one form of apparatus for
producing the effect. I prefer to employ for
this purpose an electro-magnet A Fig. 5.
having a coil upon only one of its legs (b). A
steel spring armature c is firmly clamped
by one extremity to the uncovered leg (d) of the
magnet, and its free end is allowed to project
above the pole of the covered leg. The armature c
can be set in vibration in a variety of ways – one of
which is by wind – and in vibrating it produces a
musical note of a certain definite pitch.
When the instrument A is placed in a voltaic
circuit g b e f g the armature c becomes
magnetic and the polarity of its free end is
opposed to that of the magnet underneath.
So long as the armature c remains at rest,
no effect is produced upon the voltaic current.

THE ORIGINAL VERSION OF BELL'S 1876 PATENT APPLICATION SHOWING HIS ADDITION IN THE LEFT-HAND MARGIN DESCRIBING THE CONCEPT OF VARIABLE RESISTANCE.

her father was attending business in Washington. As a result, Bell stayed at the house that night to make sure, as he put it in a letter to Gardiner Hubbard, "that burglars did not enter." While Mabel rested upstairs, Bell worked late into the night in Hubbard's mahogany-paneled library, finalizing his patent application. Bell had been toiling to complete the patent material since October, and had labored especially hard at the job over the past several weeks. He mentioned it frequently in his correspondence during the period.

January 12 was, according to Bell, a crucial night in the telephone's development because it represented his last chance to finalize his latest patent application—the one that would come to be known as the telephone patent. He had agreed to send it to Hubbard and Pollok in Washington the following day.

Because of the many irregularities surrounding Bell's patent, and the many lawsuits he would face, Bell had numerous opportunities to recount under oath his actions of that night. It was then, he says, on the eve of sending his application to Pollok & Bailey in Washington, that he realized he had inadvertently left out a crucial element—his conception of variable resistance. That is why, Bell claims, the concept is added in a lengthy note in the margin of the application and in an additional patent claim appended to the original text. As Bell explained in a 1879 deposition:

> *Almost at the last moment before sending this specification to Washington to be engrossed, I discovered that I had neglected to include in it the variable-resistance mode of producing electrical undulations.*

The following day, Bell sent the patent application to Gardiner Hubbard to give to the attorneys in Washington. A copyist in Pollok & Bailey's office reportedly drew up a "fair copy" for Bell to sign and notarize on January 20. Then, as the plan went, the application would sit, ready to be formally filed at the Patent Office as soon as Bell received word from George Brown, who was carrying it to England to file there first.

In a letter to Mabel on January 19, Bell noted that Hubbard, Pollok, and Bailey had favorably received his application and that he still needed to finalize the papers for Brown to take to England. Bell wrote:

> *I have so much copy work to do that I have employed a copyist but must still do a good deal myself—as I require three copies of each of my four specifications. One for the U.S. Patent Office, one for George Brown, and one for myself.*

The four specifications Bell refers to were all to be included in the package of patent applications he was readying for Brown, including: his first two U.S. patent filings on multiple telegraphy; his new patent application on "undulating currents" (the telephone patent); and a fourth application (never formally filed) for a device Bell called a "spark arrester," designed to suppress sparks in telegraphic and other electrical devices.

What is especially noteworthy in this passage, though, is Bell's explanation that he is, at this date, in the process of making three identical copies of each of the documents.

Five days later, on January 25, 1876, Bell, Hubbard, and Pollok met with Brown in New York City. Brown was to depart by ship to England the following day, carrying with him the package containing Bell's original, handwritten patent applications. At the meeting, Bell and Brown formalized their agreement with one another, including the provision that Bell would not file his new patent application in the United States until he had heard from Brown.

Here the mystery—already murky—deepens considerably. George Brown's copy, presumably one of the original three Bell was copying on January 19, ultimately resurfaced in 1885, when the U.S. government was preparing a case to consider annulling Bell's telephone patent on charges that it had been fraudulently won. As it turned out, Brown's copy had been retrieved by Bell's legal team years earlier, in 1878, for the *Dowd* patent case, but it had never been officially logged among that case's documents, let alone presented as evidence. The disquieting fact is this: Brown's copy of Bell's patent application contains no marginalia about variable resistance, no additional patent claim about variable resistance, and no mention of a wire immersed in liquid. All these items, so essential to Bell's ultimate success in court, are simply left out of the Brown version entirely. As a result, the version reads much like a straightforward patent application for a multiple telegraph with only the vaguest references to the possibility of transmitting vocal sounds.

When eventually confronted with this discrepancy, Bell and his legal team argued that the Brown copy that surfaced must have been an earlier draft that Bell had given to Brown months before in Canada. But the note accompanying Brown's copy when he sent it to Bell's legal team in 1878 in preparation for the *Dowd* case leaves little question about the matter. Brown states:

> *I sailed for Liverpool on the Russia from New York on Wednesday 26, January 1876—received from Professor Bell the papers I have initialed shortly before sailing. They have never been out of my possession since then until now returned to Professor Bell.*
>
> *—George Brown, Toronto, 12 November 1878.*

For reasons that have never been fully explained, Brown never did succeed in filing Bell's patents in England. Nonetheless, given the omission of any mention of variable resistance in Brown's copy of Bell's patent application, the conclusion is almost inescapable: it was missing from the version Bell handed him on January 25. Even the historian Robert Bruce, who so regularly gives Bell the benefit of the doubt in such matters of controversy, finds the omissions in the Brown patent version irregular. As he writes,

> *Whatever the reason, the fact that the George Brown specification did not mention even the principle of variable resistance suggests that the whole idea had also been left out of the American specification draft as it stood early in January 1876.*

Of course, given the fact that the Brown meeting took place *on January 25*, Bruce would have been more accurate to say that the incident suggests the idea of variable resistance had been left out up until that time. Unless Bell somehow mistakenly gave Brown an earlier, incomplete version of his application (unlikely, given Bell's diligent preparation throughout the month of January), the concept of variable resistance was simply not part of Bell's patent application at that time. If the addition had not been made by January 25, the fact casts considerable doubt on Bell's story that he added this information on the night of January 12 in a bout of inspiration "at the last moment." In other words, the language about variable resistance was most likely also absent from the version of his patent application that Bell swore to before a notary public on January 20—the version that was then purportedly held by Pollok & Bailey to ultimately be filed at the Patent Office on February 14.

IN THE COURSE of my efforts to get to the bottom of the variable resistance tangle, I happened across the fascinating work of a retired patent attorney named Burton Baker. Baker first became interested in Elisha Gray's story in 1958, while working as a patent lawyer for the Whirlpool Corporation in Michigan. At the time, the head of Whirlpool was none other than Elisha Gray II, the inventor's grandson. Upon his retirement, Baker decided to investigate the story of the patenting of the telephone in earnest, focusing on the mystery surrounding the origin of the variable resistance passage in Bell's patent application.

Baker went to great lengths to track down the story, finding no fewer than six versions of Bell's patent, including a copy of his application kept at the National Archives in Washington, D.C.; his "file copy" at the Library of Congress; and another, lesser-known copy held at the New England regional office of the National Archives in Waltham, Massachusetts. He studied everything from the handwriting on each to the type of paper the originals were written on, and he cataloged the differences among all the extant versions. Eventually, he compiled these and other findings in an interesting self-published book called *The Gray Matter: The Forgotten Story of the Telephone.*

I tracked Baker down by phone at his home in Michigan, and we compared notes at some length about the history of the telephone. Baker said that his close look at Bell's patent applications convinced him that Bell penciled in the additional information about variable resistance only after his patent application had been filed—and after he had seen Gray's caveat. As Baker states in his book:

> *It is my firm conclusion that Bell learned enough of the contents of the Gray caveat, then added the word description of a liquid transmitter to his 10-page "file copy" of the application.*

As Baker sees it, Bell's additions must have then somehow been handed to a copyist to prepare a new version that seamlessly included the additional information. The timing Baker proposes makes the prospect difficult to imagine because it means changes must have been made and recopied sometime after Bell's arrival in Washington on February 26 and yet before Wilber sent the patent to be formally printed on March 3. If Baker is right, though, Bell's central claim to variable resistance—so central to the technological and legal success of the telephone—becomes little more than a fraud.

Baker's theory involves a bit of informed speculation. What is indisputably clear, however, is that more than just the timing is awry in the fundamental mystery surrounding Bell's inclusion of the concept of variable resistance in his patent application. Bell's version of events brims with troubling inconsistencies, coincidences, and irregularities. If Bell is telling the truth, what evidence is there that he had previously considered the possibility of using variable resistance to transmit speech?

Dealing with this issue in court, Bell's legal team relied heavily upon a letter Bell wrote to Gardiner Hubbard on May 4, 1875. In it, Bell makes reference to the concept of variable resistance:

> *I have read somewhere that the resistance offered by a wire . . . is affected by the tension of the wire. If this is so, a continuous current of electricity passed through a vibrating wire should meet with a varying resistance, and hence a pulsatory action should be induced in the current.*

The letter shows that Bell possessed some basic understanding that varying the resistance of an electrical circuit could be a useful concept for his research. As he explains to Hubbard, he actually did try an abortive experiment to test this notion, using the piano of his neighbor, Don Manuel Fenollosa. But, according to Bell, the experiment failed, presumably because the piano strings were all essentially attached to one another on its metal frame, obscuring the effect. And, by Bell's own account, he never tried any other means to reduce this concept to practice prior to his telephone patent application. The key point, of course, is that the concept Bell outlines in his May 4 letter, while important, is a far cry from a workable scheme like Gray's liquid transmitter design. And Bell doesn't pursue this approach until it mysteriously turns up in the margins of his patent application some eight months later.

Even if Bell had considered the idea of variable resistance at this earlier point in his research, it is still reasonable to wonder what made him think to add it at the last minute to his patent application. Responding to a question about the matter in court many years after the fact, Bell for the first time volunteered an additional explanation. He said that in January 1876, when he was completing his telephone patent, he was also preparing to apply for a patent on his invention of a so-called spark arrester, a device designed to prevent sparks jumping from the connections of electrical devices like telegraphs with intermittent connections.

According to Bell, at the last moment before sending his telephone patent to Pollok & Bailey, he remembered his idea for the spark arrester, which involved submerging the electrical leads in water. He said the notion gave him the idea of using a liquid to vary the resistance in a circuit as a way to transmit sound. Oddly, though, in his last-minute addition, Bell didn't offer the example of using water the way he had proposed to do in the spark arrester. Instead, his patent suggests the use of mercury. As noted above, the language in the patent proposes:

> *For instance,* let mercury or some other liquid *form part of a voltaic circuit, the more deeply the conducting wire is immersed in the mercury or other liquid, the less resistance does the liquid offer to the passage of the current [emphasis added].*

If Bell was, as he attests, reminded of his water-based invention of the spark arrester, why would he suggest the use of mercury? One would imagine that Bell would have known that mercury's low resistance makes it particularly ill suited to the task he had in mind. Could it be that Bell, or whoever hastily inserted this addition to his patent, merely wanted to use an example that didn't seem so obviously cribbed from Gray's design?

Bell's explanation is instructive. He testifies that

> *This application of the spark-arrester principle to the mode of producing electrical undulations by varying the resistance of the circuit occurred to me, as I have already said, while I was making the final revision of my specification for the United States Patent Office, and almost at the last moment before sending it off to Washington to be engrossed. . . . When I came to describe the proposed mode of affecting the external resistance by the vibration of the conducting wire in water, it occurred to me that water was not*

> *a good illustrative substance to be specified in this connection, on account of decomposability by the action of the current. I therefore preferred to use as a typical example a liquid that could not be decomposed by the current, and specified mercury as the best example of such a liquid known to me.*

Of course, the vagueness of Bell's description of his variable resistance scheme involving a wire vibrating in mercury would likely have been less of a significant issue in court had the Patent Office, following its normal procedure, required Bell to submit a working model of his invention. I learned, however, that Wilber, responding to Bell's clever emphasis on his claimed invention of the "undulatory current" that made electrical transmission of sound possible, had exercised his discretion as a patent examiner in the matter, making an official notation on the file wrapper of Bell's application that no drawing or model would be necessary in Bell's case. We may never know what motivated Wilber to waive the model requirement, but the mere fact that he did so cleared up a question that had been nagging at me from the start.

Despite all the reams of testimony, it remains profoundly unclear when Bell actually inserted the language of variable resistance into his patent. It *could* have occurred at Pollok & Bailey's office prior to the application's submission on February 14, but after Pollok, Bailey, and/or Hubbard were somehow tipped off about the imminent filing of Gray's caveat. Given the tight-knit and closely connected world in which Pollok and Bailey operated, such a tip-off is very easy to imagine. A relatively small group of patent lawyers interacted regularly with the Patent Office examiners. Many of the top legal offices, including that of Pollok & Bailey, were just doors away from the handsome Patent Office building near the Capitol. A small network of regular inventors, draftsmen, and copyists all worked closely with one another and no doubt socialized, making the exchange of information about the latest new invention commonplace.

It is even possible to imagine, given the circumstances involved, that

Bell himself did not make, authorize, or even know about the addition, just as he claims not to have known about Hubbard's abrupt decision to file that day. Whatever information Hubbard, Pollok, Bailey, or someone else from their office could have had about Gray's imminent filing might have been hastily used to add language to defend Bell's claim against Gray's caveat. In such a scenario, Bell might have later put the additions in the margins of his own file copy to make it fully comport with the one that had been filed.

Information about variable resistance could also have been added after Bell's meeting with Wilber in Washington. Wilber does not confess to having allowed such a change; however, he does comment that security at the Patent Office was lax, and it would certainly have been possible, given Pollok and Bailey's connections, for someone to have removed Bell's patent and made changes to it after hours.

However it might have happened, though, the *facts* of the case strongly suggest that the addition of the concept of variable resistance to Bell's patent was made after the rest of his application had been completed. Furthermore, the text Bell inserted into his patent bears such a striking resemblance to Gray's description in his caveat that it cannot be easily explained away.

A final inconsistency to consider is this: If, as Bell says, Wilber's only hint about Gray's work came by pointing to the paragraph inserted in Bell's application concerning the use of "mercury or some other liquid" to vary resistance, why would Bell write to Gray in 1877 that he had heard that Gray's caveat concerned a wire "vibrating in water"?

Amazingly enough, in the voluminous Bell archive, even this question was directed to Bell in one of the many court cases in which he testified. As the opposing counsel asked Bell:

> *How did you come to know that Elisha Gray's caveat "had something to do with the vibration of a wire in water" and just what was the full extent of your information on the subject, up to the date of your said letter of March 2, 1877?*

In response, Bell seems to be uncharacteristically at a loss for an explanation:

> *I do not know what it was that led me to imagine that the caveat had something to do with the vibration of a wire in water, unless, indeed, it might have been due to the fact that the liquid I had in contemplation when I first wrote the paragraph from my specification quoted above, was water, and not mercury, as stated in that paragraph; and to the fact that I had previously experimented with water in my spark arrester.*

Among the many close observers of the Bell saga to note the flimsiness of this explanation is Lloyd Taylor. As he aptly puts it,

> *On this gossamer thread, spun four years after the event, hangs Bell's sole claim to priority in the concept of the liquid transmitter.*

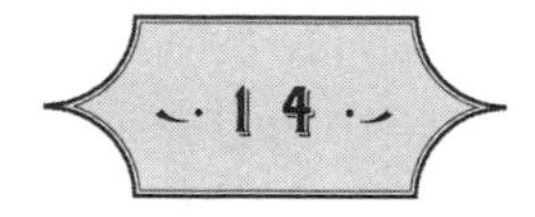

CALL WAITING

I HAD UNEARTHED enough information so far to raise serious doubts about Bell as the sole inventor of the telephone. And yet, over a century later, this is largely how he is remembered. The image of Bell as the smiling, portly, white-haired "father of the telephone" lives on today in science textbooks, children's stories, and scholarly works alike. I was not entirely sure how the myth of Bell's single-handed invention of the telephone managed to survive unscathed in his own time in the face of the incriminating evidence against him that emerged in court—from the irregularities at the U.S. Patent Office to the suspiciously missing paragraph about variable resistance in the version of his patent application Bell sent to England. But I was even more mystified by the way Bell's stature in the public imagination could have proved so unshakable given a steady onslaught of efforts by assorted researchers, including Silvanus Thompson, Lloyd Taylor, Burton Baker, A. Edward Evenson, and many others, over the course of more than a century to correct the record.

Clearly, one explanation is that history's winners stand in the best position to control the way events are remembered. And the Bell Telephone Company, which began to amass immense profits from its monopoly position as the 1800s wore on, was certainly a winner. For a dramatic example of this phenomenon at work in the case of the telephone, consider the case of George Prescott.

In 1878, Prescott published a book entitled *The Speaking Telephone, Talking Phonograph and Other Novelties.* Prescott was a respected electrical researcher and author and a contemporary of Gray and Bell. In 1878, he served as Western Union's chief engineer, which meant that he would have likely tended to ally himself more closely with Gray's work than Bell's. By then, Gray was working as an independent inventor, but he still licensed many of his inventions to Western Union and its subsidiary Western Electric. Despite Prescott's corporate affiliation, though, he was also a prolific writer with a reputation for fairness and independence in his work. He was also certainly one of the most knowledgeable individuals of his day to report on the developments surrounding the introduction of the telephone.

In the first edition of Prescott's book, written two years after Bell received his patent on the telephone, Prescott explicitly credits Gray with the invention. After reviewing the features of all the brand-new telephone models of his day, including Bell's, he notes that

> *all the Speaking Telephones which we have described possess certain common characteristics embodied in Mr. Gray's original discovery, and are essentially the same in principle although differing somewhat in matters of detail.*

As it turns out, Prescott's first edition is now exceedingly rare. When Western Union settled its *Dowd* lawsuit with the Bell Telephone Company in November 1879, the terms of the settlement dictated that Western Union had to unequivocally acknowledge Bell as the inventor of the telephone. Prescott, given his affiliation with Western Union, may have felt obligated to accede to the demand. One way or the other, however, the history of the telephone presented in his book was thoroughly rewritten in the subsequent 1884 edition. Consider the passage cited above, for instance. It was changed to read as follows:

> *all the Speaking Telephones which we have described possess certain common characteristics embodied in* Mr. Bell's original discovery, *and are essentially the same in principle although differing somewhat in matters of detail [emphasis added].*

Perhaps an even more revealing change was made to Prescott's particular discussion of the liquid transmitter. In his first edition, Prescott presents a diagram of the liquid transmitter (labeled fig. 52) accompanied by the following note at the bottom of the page:

> *From the reading of the text it might be erroneously inferred that the apparatus shown in figure 52 was invented by Professor Bell, and exhibited by him at the Centennial Exhibition. Professor Bell neither invented nor exhibited it. The above figure represents the transmitting portion of Elisha Gray's original Speaking Telephone—the first articulating telephone ever invented.*

Prescott's analysis in his first edition could hardly be more explicit. But, of course, this informative note is omitted entirely in subsequent editions. Fittingly, even the title of his book was changed. From 1884 on, it was published as *Bell's Electric Speaking Telephone: Its Invention, Construction, Application, Modification and History.* Equally important, this whitewashed version of Prescott's initial work would become a significant and widely read text, published most recently in a 1972 reprint by a former publishing subsidiary of the *New York Times*. As the saying goes, to the victor go the spoils.

Prescott's case, fully exposed by previous writers of telephone history including Lloyd Taylor and A. Edward Evenson, is of course an extreme case. To be sure, many more subtle pressures must be at work to explain why certain kinds of erroneous historical myths seem to stubbornly survive all efforts to amend them.

Over the course of the year, I discussed the issue of historical myths on a number of occasions with George Smith, the director of the Dibner Institute, who was particularly interested in the topic. One day, he dropped by my office to share a draft of a preliminary proposal he had drawn up in discussions with a public television station for a program tentatively entitled "Myth versus Reality in the History of Science." The proposal featured a number of examples, but one particularly caught my eye.

Drawing specifically upon his expertise about the life and work of Sir Isaac Newton, Smith used the example of the myth which suggests that Newton's theory of gravity was sparked by an apple falling in his mother's garden. In one common version of the tale, believed to have originated in an account in 1839, more than 150 years after the fact, the apple story is embellished. In this still commonly repeated version, the apple didn't just idly inspire Newton's imagination, but even "struck him a smart blow on the head."

Smith explained emphatically that there is absolutely no evidence in the historical record to suggest that Newton got his inspiration from an apple, not to mention getting hit by one. According to the myth, the event was supposed to have occurred some twenty years prior to Newton's publication of his theory of gravity. But historians like Smith who have studied Newton's papers and correspondence generally agree on the sequence of steps that led to Newton's conception of gravity, and it involves no reference to falling fruit of any kind. Nonetheless, as Smith writes in his proposal,

> *The apple myth has become part of Western culture. The myth that Newton had his theory of gravity twenty years before publishing it continues to appear in textbooks, including a recent otherwise good text on general relativity.*

Smith said he suspected that, when it came to topics like Newton's apple, science textbook authors deserved a good deal of blame. "Textbook authors tend to put great effort into avoiding falsehoods in their portrayal of science," he said, "but they often seem to simply recycle their historical remarks from other textbooks without bothering to check them against actual historical texts."

SMITH'S STORY ABOUT the persistence of Newton's apple gave me an idea for how I might retrace the way the story of the telephone has been told over the years. Tapping the extraordinary and still-emerging power of full-text book searching offered by Google and Amazon.com, I decided to search for incidences of the phrase "Watson, come here." That way, I figured, I could quickly locate a good sampling of the published iterations of the well-known tale of the telephone's invention.

Within seconds, my initial Google search yielded more than three hundred results, culling to my computer screen passages from books of every kind—from electronics textbooks and obsolete technical encyclopedias to children's stories. Aside from a few stray passages from novels—including some by Sir Arthur Conan Doyle—that contained a character named Watson, the results offered a kaleidoscopic collection of the way Bell's "eureka moment" has been disseminated in print. Clicking through these entries, I could quickly skim the pertinent passages from many of these books.

The first thing I noticed is that the iconic tale of Bell calling to Watson is often told in such a stripped-down, abbreviated way as to render it all but meaningless. Consider, as a prime example, Irving Fang's textbook, *A History of Mass Communication* (1997), which notes:

> *On March 10, 1876, Bell said to his assistant from an adjoining room, "Mr. Watson, come here. I want you," and the telephone was born.*

Passages like this can be taken as nothing more than an empty referent, unless we are to believe that Bell's phrase somehow conjured the telephone into being. Tipping his hat to Bell, the author simply assumes we already know all about its invention.

Beyond these frequent passing mentions, the next surprise was the number of obvious errors that crept into the "Watson, come here" narrative. A wonderfully pretentious volume, *The Nobel Book of Answers* (2003), offers Nobel laureates' answers to "some of life's most intriguing questions." Gerd Binning, who won the Nobel Prize in physics in 1986, does a serviceable job answering the question: "How Does the Telephone Work?"—with the exception of this rather stunning vignette:

> *According to popular accounts, the long-awaited breakthrough came only by coincidence: Bell's assistant, Watson, had spilled some acid in the lab next door. In his distress he yelled, "Mr. Bell, come quickly!" Seconds later the door opened and Bell came hurrying in—he had heard his colleague, not through the wall, but through the experimental device that connected both rooms.*

Binning's error—describing Watson as the one doing the calling—was no doubt a simple mistake made, just as George Smith had described, because the author hadn't bothered to double-check the history. More often, though, the errors were subtler, revealing deeper truths or points of controversy. For instance, a surprising number of texts got the date wrong. Ian McNeil, for example, writing in the *Encyclopedia of the History of Technology* (published by Routledge in 1996), recounts that

> On 7 March 1876 *[emphasis added] the famous command, "Mr. Watson, come here. I want you!" was uttered by Bell to his assistant, who instead of only hearing Bell's voice from the other room, heard it over the primitive induction device with which they were experimenting.*

Curiously, the same incorrect date for this incident (March 7, as opposed to March 10) appeared in at least a half-dozen texts I reviewed, such as Scholastic Books' *Famous Americans: Twenty-two Short Plays for the Classroom* (1995), in the form of a theatrical skit to be read aloud by elementary school students. Act 2, Scene 1 of the skit about Bell is presented as follows:

> ***Narrator:*** On March 7 of the following year *[1876], [emphasis added] Bell and Watson were experimenting in separate rooms, trying out a new transmitter. Suddenly, Bell spilled battery acid on his clothing.*
> ***Bell:*** *Mr. Watson, come here! I want you!*
> ***Watson (rushes into room):*** *Mr. Bell! It works! I heard your voice perfectly over the wire! You said, "Mr. Watson, come here! I want you!"*
> ***Bell:*** *So human speech can travel by wire after all! My telephone works!*

What makes these particular errors consequential is that March 7, 1876, is the day that the U.S. Patent Office officially granted Bell's telephone patent. The texts that contain this particular error all neatly obliterate the dissonant fact that Bell had yet to successfully transmit intelligible speech on the day he received his patent.

The location of the Bell and Watson experiments is also frequently mistaken, and not just because, as I had learned on my Boston sojourn, the cityscape has changed. Rather, I traced the origin of this family of errors at least as far back as a widely read book from 1910 by Herbert Casson, *The History of the Telephone*. In his account, Casson forgets (or perhaps never knew) that, by the time of the "Watson come here" moment, Bell and Watson had moved their operations from the Williams shop at 109 Court Street where, upon occasion, Watson had tried to listen to Bell's telegraph prototypes from the basement while Bell toiled in the attic. Of course, we know from Bell and Watson's accounts that the scene occurred in adjoining rooms in Bell's newer laboratory, in a boardinghouse at 5 Exeter Place. Casson, however, describes it this way:

> *"Mr. Watson, come here, I want you." Watson, who was at the lower end of the wire, in the basement, dropped the receiver and rushed with joy up three flights of stairs to tell the glad tidings to Bell. "I can hear you!" he shouted breathlessly. "I can hear the words."*

Casson's error, introduced nearly a century ago, seems to have spawned confusion that lives on in many modern texts. The fourth edition (2001) of the textbook *Understanding Telephone Electronics*, for example, explicitly places the scene at 109 Court Street in Boston. So do at least four other texts I found. Many others leave out the location but retain the erroneous "upstairs-downstairs" element of the story. Consider, for instance, how *Alexander Graham Bell (On My Own Biographies)*, a 2001 children's book about Bell for beginning readers, preserves Casson's confusion:

> *A pile of tools, wires, and battery acid surrounded him. Suddenly, Aleck spilled some acid. The acid burned. Aleck cried out, "Mr. Watson, come here! I want to see you." He forgot that Watson was too far away to hear him. To Aleck's surprise, Watson came running upstairs. He burst into the room. "Mr. Bell, I heard you!" he said. "I heard every word!"*

Were all historical accounts riddled with such problems, I mused. If historians and writers of every stripe could not agree on the place, date, or circumstances of a well-known and well-documented incident like the telephone's invention, how much credence should we give to the standard renditions of any historical tale?

I was so busy puzzling over the odd inconsistencies in this story that I nearly overlooked something far more consequential: Casson's bungled 1910 account was the earliest version of the famous scene that showed up in my search.

Looking further, I realized that, however unlikely it seemed, there were no contemporaneous versions of this story, no speeches about it, no newspaper accounts from the 1870s. Scouring all my sources, I discovered that Bell himself never publicly told this now-famous story. The discovery seemed hard to believe until I realized that dreaded Whiggism had crept into my analysis. Just as David Cahan had warned, I was reading history backwards. The scene is so closely associated with the telephone today, I found it hard to accept that it was virtually unknown at the time. After all, it does purport to recount the world's first conversation over the telephone.

By all accounts, Bell loved nothing more than holding forth about his scientific research, and this was certainly his greatest breakthrough. It was hard to imagine that Bell would never have regaled people with the story of his first telephone call. Yet, as I concluded after a careful review, the story we now quote would probably be all but unknown today if it hadn't been authoritatively and winningly recounted by Watson in his autobiography, *Exploring Life,* published in 1926—four years after Bell's death.

The more I thought about it, the more consequential Bell's silence seemed to be. In the story as Watson eventually recounts it, Bell calls out urgently after *spilling acid on his pants.* Even the basic storyline evokes Bell's work with a liquid transmitter. Did Bell withhold any public mention of his success with a liquid transmitter to keep Elisha Gray in the dark about his foul play? Was that why he never told the story of his first success in transmitting intelligible speech? I wasn't sure yet, but it certainly seemed like a plausible explanation.

The explanation gained further credence when I learned, to my amazement, that in the 1879 *Dowd* case—the only occasion when the claims of Bell and Gray came directly into conflict in court—Bell made no mention of his first transmission of speech with the liquid transmitter *during his entire nine days* on the witness stand.

The following year, in 1880, Bell does touch upon the story under questioning in the courtroom. And Watson gives a somewhat more detailed account during his testimony in August 1882. But aside from Casson's flawed telling, the story does not come to widespread public attention until some *four decades* after the fact, when Watson recounts it in his autobiography. Only then does it begin to make its way into the history books, starting with Catherine MacKenzie's *Alexander Graham Bell: The Man Who Contracted Space,* the first biography of Bell, published in 1928. MacKenzie's book drew upon her close association with Bell as his assistant during the last eight years of his life. Despite Watson's prior published account, one wonders about the extent to which MacKenzie's rendition posthumously derives from Bell himself. As she tells it:

> *Watson dashed down the hall into the laboratory. Bell had upset the acid of a battery over his clothes. In his delight over Watson's sudden appearance, Bell forgot all about the spreading acid stains on his trousers and flew to the other end of the wire, to hear Watson's voice now coming clearly through.*

ON A PARTICULARLY cold and wintry afternoon, I was heading out to have lunch in Cambridge with a colleague who taught in MIT's graduate writing program. As I often did, I walked through the main reception area of the Dibner Institute to check my mailbox. My thoughts were occupied with the events of 1876 and Bell's strange silence about his telephone breakthrough. On my way through the front office, I passed the handsome glass cases containing the assortment of familiar historical artifacts, when one of them suddenly grabbed my attention in a new way.

There, behind the glass, rested the handsomely marbled cover of a pamphlet from the American Academy of Arts and Sciences. Dated

May 22, 1876, it was a transcription of Alexander Graham Bell's first public speech about the telephone. I had walked past it countless times without giving it much thought, but now it beckoned with a crucial clue. I realized that in a year of research I had been so focused on the details leading up to the invention of the telephone that I had overlooked Bell's actions afterward. I had seen many references to Bell's talk at the Academy and read excerpts of it, but I had never closely inspected the speech itself. Given Bell's reticence about the "Watson come here" story, how exactly had he recounted the circumstances of his invention of the telephone when he first spoke of it in public?

Before heading off to lunch, I sped to my office to look online in the hope that I could locate a copy of Bell's speech at one of MIT's libraries. Sure enough, MIT's Hayden Library in the heart of the campus contained a full set of the *Proceedings of the American Academy of Arts and Sciences*, dating back to the organization's founding in 1780.

I made for the library directly after my lunch meeting.

The works I sought were stored in a section in the basement, where the floor-to-ceiling shelves roll on tracks to close up solidly against one another, packing the walls of books and journals together like proverbial sardines. To retrieve a book, the intrepid researcher has to turn a crank, roll apart the walls of bookshelves, and create a passable aisle, then brave this new crevice to find what you are looking for.

The musty old volume from 1876 looked as if it hadn't been touched for many decades. Making my way straight to the nearest carrel, I sat in the fluorescent glare to examine it.

Scholars normally describe Bell's Academy speech as the first time he publicly disclosed his work on the telephone. And yet, looking back over the bound volume of the Academy's *Proceedings,* I realized it was doubtful that Bell had actually demonstrated the transmission of speech at this oft-cited event. Without offering many specifics, Bell did mention that he had transmitted some intelligible speech with his invention. But he did not reveal the circumstances of his initial success, and he made only a passing conceptual reference to the liquid transmitter. Instead, Bell focused on, and presumably demonstrated for his colleagues, a far more primitive apparatus he had built that could transmit only musical tones.

Once again, I marveled that the truth seemed so at odds with the received history. Hot on this new trail, sitting in my office that evening, I looked further into the matter using the online collection of the Library of Congress. I quickly found a letter Bell had written to his parents immediately after giving his presentation at the Academy. It corroborated my suspicion:

> *The meeting at the Academy was a grand success. I had a telegraph wire from my rooms in Beacon Street to the Athenaeum building and my telegraphic organ was placed in my green reception room under the care of [Mabel's twenty-six-year-old cousin] Willie Hubbard.*

Bell wrote that he telegraphed Willie to play some music, "and in response came some rich chords." But he made no mention of having tried to speak through his device. Sometime later I found that, as was so often the case, Bell's legal testimony offered the most precise and revealing account. The following passage from Bell's deposition in the 1879 *Dowd* case would seem to leave little doubt about the matter. The opposing counsel specifically asks Bell what electrical devices, if any, Bell exhibited during his speech at the American Academy. Bell responds:

> *I do not know that I can recall them all, but I remember some. I exhibited the membrane telephones referred to in section 12 of the paper. I also exhibited what was called, in the former telephone cases, the "iron-box" receiver. I also showed, I think, circuit-breaking transmitters and tuned-reed receivers, and I think, also showed a liquid speaking telephone transmitter, like that referred to in section 13 of the paper, but I am not quite certain of this.* If it was shown, as I think it was, it was not shown in operation *[emphasis added].*

Despite his marked and uncharacteristic vagueness in this answer, Bell seems notably certain about that last point. It is truly a remarkable admission. Consider the fact that Bell had made his breakthrough speaking to Watson with the liquid transmitter some two months earlier. It is possible to imagine that Bell might have perhaps been concerned that the liquid transmitter might not perform reliably and thus refrained from demonstrating it. But such a hypothesis seems unlikely. Bell had built a device capable of transmitting speech. And he had a U.S. patent giving him broad rights to it. His speech to his prestigious colleagues at the American Academy marked his moment to bask in their acclaim. Why, then, would he refrain from displaying his most successful and important breakthrough?

Reviewing Bell's laboratory notebook once again, I realized that, during this period in the early spring of 1876, after Bell's phenomenal success with the liquid transmitter, he quickly abandoned any effort to improve his liquid transmitter design. Instead, he switched his focus to developing *an alternative method* to transmit speech that evolved directly from his accident almost a year earlier in which Bell had heard Watson pluck the steel reed of his finicky multiple telegraph receiver.

In so doing, Bell switched to trying to develop a "magneto-electric" transmitter design. He had guessed from his earlier work that such a design would be possible. But, at the time he filed his patent, he provided no specific guidelines for how to build such a transmitter and, of course, he himself had never yet done so to successfully transmit articulate speech.

Like the liquid transmitter, Bell's magneto transmitter design used a vibrating diaphragm, but it operated on a different principle. Rather than using the vibrating diaphragm to vary the electrical resistance in the circuit, the diaphragm vibrated in front of a magnet to cause tiny fluctuations of the current in the magnetic field. It was, in essence, a telephone receiver in reverse: rather than using fluctuations in electric current to trigger vibrations of sound waves in the air, Bell realized he could make the sound waves induce tiny changes in a magnet's residual electric current.

The method yielded a terribly weak signal, but Bell nonetheless had adopted the method wholeheartedly by late April 1876. By May, all mention of the liquid transmitter in Bell's laboratory notebook had ended. Only then, with this alternative transmission method in hand, did Bell finally make a public announcement of his invention of the telephone.

In 1966, Bernard Finn, a curator at the Smithsonian Institution, tested some of the museum's collection of Bell's transmitters and surmised that Bell might have switched away from the liquid variable resistance transmitter because the magneto design worked better. Finn's hypothesis is intriguing, but Bell's notebooks offer no evidence that he was displeased with the performance of the liquid transmitter, or that his new method achieved any improvement. On the contrary, Bell notes that it is difficult to hear consonants with the early versions of his magneto transmitter.

Even if Finn's conjecture is accurate, it still does not adequately explain Bell's complete abandonment of the variable resistance transmitter. Bell surely recognized that the signal in his magneto transmitter was exceedingly weak, whereas the variable resistance transmitter's signal could be easily amplified. This crucial advantage would make it possible to transmit speech over long distances. Electrical researchers, including most notably Thomas Edison, would soon make the need for liquid obsolete by using carbon in their variable resistance transmitter designs. But it was the liquid, variable resistance transmitter which first crossed the vital technological threshold, opening the door to virtually all modern telephone transmitter designs.

Viewed with hindsight, the irony is that Bell took a marked step backward in his transmitter design, moving from a primitive version of the variable resistance model that would become the industry standard to a magneto-electric design that would quickly become obsolete. As Bell's biographer Bruce notes,

> *Upon the variable-resistance transmitter, drawing power from a readily expansible source, the future telephone industry would*

> *rest. Bell saw its great advantage clearly enough. Yet by the end of April [1876] he had drifted away from it and back to the magneto transmitter, which depended on the puny power of the sound waves themselves to induce its current.*

Nonetheless, having finally succeeded in transmitting speech—months after the fact—with a telephone design that comported more closely with the language in his patent, Bell went public. He announced his success in his lecture at the American Academy. And he actually demonstrated speech transmission several days later, on May 25, 1876, before a good-sized audience at MIT.

With help from the staff of the MIT Archives, I retrieved the handwritten minutes from the MIT meeting at which Bell appeared. This time, unlike his speech at the American Academy, Bell demonstrated the speaking telephone in action. A brief report in the *Boston Transcript* notes that vowel sounds came through Bell's telephone device, but that consonants were all but "unrecognizable." Notably, at this landmark event, Bell demonstrated only his magneto-electric transmitter; the liquid, variable resistance transmitter was nowhere to be seen.

Once again, the evidence, while circumstantial, all pointed toward one conclusion. On March 10, 1876, Bell crossed a momentous threshold and he knew it immediately, writing to his father on the night of his success that he had "at last found the solution of a great problem." Plus, he already held a patent protecting his invention. In public, however, Bell made *absolutely no mention* of the circumstances surrounding his initial success. Then, when Bell did finally unveil his speaking telephone publicly, he demonstrated only his second, and likely inferior, transmitter model. Why? A close look at Bell's actions strongly suggests that he sought to conceal the fact that his telephone breakthrough came using a liquid transmitter design. Under the circumstances, there is little doubt that Bell wished to hide this incriminating information—especially from Elisha Gray.

It seems likely that, despite Bell's success in March with the liquid transmitter, he waited to go public until he had found an alternative transmission method he could reasonably claim as his own. Bell no doubt hoped that, by featuring the magneto transmitter, he could present a plausible and seemingly independent path to the telephone that would dispel any hint of foul play. If so, Bell's plan would be put to a particularly nerve-racking test on June 25, 1876. On that date, Bell had to demonstrate his telephone and explain his research to an international group of telegraph experts and dignitaries that included Gray himself.

PARTY LINE

LATE INTO THE night on Wednesday, June 21, 1876, Alexander Graham Bell sat awake in a gaslit room at Philadelphia's Grand Villa Hotel, writing a long letter to his fiancée, Mabel Hubbard. He was too excited and nervous to sleep. Bell had traveled to Philadelphia to exhibit his telephone at the International Centennial Exposition, an enormous world's fair held to commemorate the nation's one hundredth anniversary. Shortly after his arrival, Bell told Mabel, he had learned that the Centennial's eminent panel of judges—including Britain's Sir William Thomson (later Lord Kelvin), one of the world's most respected electrical researchers—expected Bell to demonstrate his invention that coming Sunday.

A month earlier, Bell had given two well-received talks about his research to rarefied academic audiences in Boston, but the stakes here in Philadelphia, amid the fanfare of the exposition, were far higher. Hubbard, ever the enterprising businessman, had insisted that the reluctant Bell unveil the telephone to the public at the exposition because of the event's size and global audience.

THE OPENING CEREMONY OF THE CENTENNIAL EXPOSITION IN PHILADELPHIA, MAY 10, 1876, PRESIDED OVER BY PRESIDENT ULYSSES S. GRANT AND EMPEROR DOM PEDRO II OF BRAZIL.

On this point, Hubbard's judgment was astute. The Centennial Exposition was truly a grand venue for such a technological debut, affording a level of visibility virtually unobtainable elsewhere. More than 30,000 exhibitors were displaying their wares in 190 buildings on the 250-acre site in Philadelphia's vast Fairmount Park. Over the course of its six-month run, the exposition would draw an astonishing 10 million visitors from around the world.

Bell had arrived in Philadelphia two days earlier, on June 19, and, like the throngs of other visitors, he had marveled at the exhibition's offerings from the moment he visited the fairgrounds. As he wrote Mabel:

Visitors to the Centennial could climb to the top of Frédéric Bartholdi's "arm and torch" statue, a work-in-progress that would eventually become the Statue of Liberty.

I really wish you could be here, May, to see the exhibition. It is wonderful! You can have no idea of it till you see it. It grows upon one. It is so prodigious and so wonderful that it absolutely staggers one to realize what the word "Centennial Exhibition" means. Just think of having the products of all nations condensed into a few acres of buildings.

Some thirty-seven nations had sent examples of their handiwork, from European paintings and sculpture to Chinese porcelain and carved jade. One of Brazil's contributions, an ornate, solid-silver, two-ton table, particularly caught Bell's eye. Venezuela's exhibit featured a portrait of George Washington that had purportedly been woven from the hair of Simon Bolívar, the Venezuelan revolutionary leader who was revered as a liberator throughout South America. Prominent among such international offerings was the French sculptor Frédéric Bartholdi's dramatic, unfinished statue of a huge hand bearing a torch. This massive work-in-progress, several stories tall and hand-fashioned from copper, stood outside with its own small pavilion near the center of the exposition. For a small charge, visitors could walk into the statue and climb to the base of the torch. They were also asked to donate ten cents apiece to allow Bartholdi to complete his wildly ambitious project. It would take another decade, but the arm and torch would eventually sit atop Bartholdi's monumental Statue of Liberty, an ode to America that found its permanent home in New York Harbor in 1886.

MORE THAN ANYTHING, though, the Centennial Exhibition was a showcase for the burgeoning age of invention and, in particular, for American machinery and ingenuity. The Centennial boasted the world's first steam-driven monorail system and even a newfangled elevator on Fairmount Park's Belmont Hill—it could carry forty people at a time to the top of a 185-foot-tall observation building for a bird's-eye view of the grounds.

In the vast Machinery Hall, the latest technology was on display in all its Gilded Age splendor. Visitors most likely got their first-ever glimpses of sewing machines and typewriters, not to mention a stag-

Visitors to the Centennial Exposition's Machinery Hall were dwarfed by the 56-ton flywheel of the Corliss Steam Engine on display.

gering array of projection lanterns, fire alarms, lathes, and firearms. A vast row of locomotive engines on display gave way to a colossal 70-foot-tall, 1,400-horsepower Corliss Steam Engine whose 56-ton flywheel commanded the hall. The giant steam engine powered all the exhibits in the building through five miles of overhead belts, shafts, and pulleys. Even beyond its impressive function, though, the engine stood as a potent symbol for the entire exhibition. As one correspondent colorfully noted, it rose

> *loftily in the center of the huge structure, an athlete of steel and iron with not a superfluous ounce of metal on it; the mighty walking-beams plunge their pistons downward, the enormous fly-wheel revolves with a hoarded power that makes all tremble.*

It was no wonder that inventors, craftsmen, and representatives from leading manufacturing companies of all kinds came to display their products. Edison brought his newly designed duplex telegraph, able to send two messages at once. Rudolph Koenig of Paris, whose name I remembered thanks to my colleague Dave Pantalony's special interest in him, had set up an elaborate exhibit to display his unrivaled scientific tuning forks and other acoustic devices. The New York firm of Bausch & Lomb offered an impressive display of optics, from eyeglasses to hand telescopes and binoculars. Another prominent exhibit touted the extraordinary heat resistance of a brand-new material called asbestos. And, in the fair's restaurant section, a Pittsburgh pickle merchant named Henry Heinz introduced the public to a new tomato condiment.

Bell must have recognized that the Centennial represented an unmatched opportunity to display the fruits of his research, yet his correspondence reveals a person racked with trepidation as his moment in the spotlight drew near. As Bell curiously complained to Mabel, for instance:

> *I shall be glad when the whole thing is over. I wish the whole telegraph were off my hands altogether.*

Bell's tone is certainly not what one would expect from someone who, with a U.S. patent firmly in hand, was about to unveil a remarkable and unprecedented invention to the world. While it is understandable that Bell would be nervous before his performance, his hand-wringing comes across as more than simply a matter of last-minute jitters.

Bell had resisted the idea of displaying the telephone at the Centennial Exposition ever since Gardiner Hubbard had suggested it months earlier. First, Bell missed the application deadline; Hubbard, undaunted, used his influence with the Centennial's organizing committee to allow Bell to participate anyway. Then, Bell continued to refuse to go to Philadelphia; but Hubbard made his case to Mabel, who took the matter into her own hands.

Ultimately, in an episode that would become part of Bell family lore, he left for Philadelphia under the most remarkable circumstances. As Bell's daughter Elsie recounted much later, Mabel, after much fruitless cajoling to get Bell to attend the event, finally resorted to a ruse: she persuaded him to accompany her on an innocuous-sounding "drive in the family carriage," only to whisk him to the train station and hand him a suitcase that she had surreptitiously packed with his belongings.

Bell, stunned, still declined to get on the train, but Mabel burst into tears and threatened not to marry him if he didn't go. As Elsie tells the tale, Mabel (who was of course deaf) turned her head away so she couldn't hear Bell's protests in retort. In his own version of events, Bell explains in a letter to his mother from New York the next day that he grudgingly boarded the train at the last possible moment because he couldn't bear to see Mabel so "pale and anxious."

The incident explains why Bell wrote Mabel upon his arrival in Philadelphia:

> *My darling May:*
> *Fate has brought me here against my will and now she seems determined to keep me in spite of myself. . . . [I] feel that I am here for your sake and at your device.*

For her part, Mabel acknowledged that the episode had been hard on her, too. As she put it:

> *It was very hard to send you off so unwillingly but I am sure it was for the best and you will be glad of it by and by. Don't get discouraged now. If you persevere, success must come. Anyway it will be a great help to you to be connected with scientific men.*

Still, the circumstances explain how Bell got to Philadelphia against his will, but not why he was so set against the idea in the first place. Most of Bell's biographers, taking his protestations literally, have explained his reluctance to exhibit his telephone in Philadelphia as a consequence of his dedication as a teacher, noting that it was an inconvenient time for him to leave his students. True, it was near the end of the semester, but, as it turned out, Bell was back in time for final exams. However dedicated he might have been to his students, it is difficult to imagine that leaving them for a few days would have caused him such consternation.

Similarly, Catherine MacKenzie suggests that Bell's reluctance stemmed from the fact that he had missed the deadline for entering his device into the exhibition and did not want Gardiner Hubbard, who served as part of the Massachusetts delegation to the Centennial, to have to use any influence on his behalf. As MacKenzie recounts,

> *Mr. Hubbard assured him that it would be the merest detail to add the apparatus to the earlier exhibit. In his official capacity nothing could be easier to arrange. To Bell, whose literal honesty never made concessions, the thing was impossible. If he could not comply with the rules, he would make no exhibit.*

This, too, seems like an inadequate explanation. Even a cursory look at the emotional tone of the correspondence during Bell's visit to Philadelphia suggests that considerably more was going on.

Upon closer inspection, it seems apparent from Bell's many references in letters to his parents, and to Mabel, that a major cause of concern was the presence at the exposition of none other than Elisha Gray, who had come to demonstrate his own multiple-messaging telegraph at the prominent Western Union exhibit. Although the two men had never met, Bell kept close tabs on the inventor he considered his rival and surely knew that Gray had been in Philadelphia for over a month readying his exhibit.

Bell's concern appears to have reached a level of near panic when the Centennial's judges asked him to demonstrate his devices on Sunday *immediately following their review of Gray's instruments.* The prospect of such a juxtaposition realized Bell's worst fears. As he explained to Mabel in his confessional, late night letter from the Grand Villa Hotel:

> *I must say I don't like this at all. . . . I feel very nervous about it—for I feel I have come on here very hurriedly without sufficient preparation* to be thrown into direct collision with Mr. Gray *[emphasis added].*

Although Bell would often worry, his level of apprehension and foreboding, like his reluctance to attend the exposition in the first place, stands glaringly at odds with the exciting promise of his circumstances. Bell was a natural orator and performer. It is hard to interpret anything other than guilt and panic in his dark brooding on the eve of his Centennial demonstration. Why else would he write to Mabel that he felt "very hopeless" and "heartbroken" just as he was about to showcase the greatest success of his life? Bell even frets to Mabel that

> *If I don't make a change and very soon—Telegraphy and Visible Speech together will be the end of me—and then we shall never be married at all.*

What kind of change was Bell seeking to make? And what exactly was he saying would lead to his demise? The florid hand-wringing in his correspondence does not let on. But Bell's mood is revealing, and so are his actions. He surely wanted to impress the judges with his demonstration any way he could. And, to that end, the telegraphic equipment sent to Philadelphia on Bell's behalf included the liquid transmitter that he had used with such success when he first transmitted speech. With Gray present, though, how could Bell display his transmitter, or even reveal a hint of its existence? Little wonder Bell was nervous: he faced not one but two daunting tasks. He had to display a working telephone to the judges *and* convince Elisha Gray that he had invented the telephone independently. Only then could he hope to walk away from the exposition with his reputation, his sole claim to the telephone, and his engagement intact.

AFTER BELL HAD prepared and made adjustments for several days, the dreaded Sunday, June 25, arrived. The judges had chosen that day to review Gray's and Bell's telegraphic equipment because the exposition was closed to the public, affording relative quiet for the acoustic demonstrations. The morning proved to be not just quiet—it was also swelteringly hot, especially inside the glass pavilion of the main hall where the formally attired delegation of fifty scientists and dignitaries gathered. Among them, only the gregarious and rotund Dom Pedro II seemed relatively unruffled by the heat.

At the Western Union exhibit on the center aisle of the main hall, Elisha Gray gave an impressive demonstration of his harmonic telegraph, which simultaneously transmitted a remarkable eight messages over one wire. More comfortable as an engineer than an orator, Gray had brought with him Professor George F. Barker from the University of Pennsylvania to offer additional theoretical explanations for his devices. To great acclaim from the judges, Gray also demonstrated his musical telephone. Transmitting from several hundred feet away, Gray played a rendition of "Home, Sweet Home," which came through clearly even over the gasps of excitement and amazement from the distinguished audience.

Finally, it was Bell's turn. Gray's demonstration had been as polished and professional as Bell had feared; but Bell, unlike Gray, was a skilled orator. He knew he could captivate the judges—if his speaking telephone would operate successfully. Bell didn't have a professor of George Barker's stature to assist him, but he did have Mabel's cousin, Willie Hubbard, who had helped with the demonstration at the American Academy of Arts and Sciences. He had come to Philadelphia with his uncle Gardiner for the event. Bell also had a stroke of luck in having recently met His Majesty Dom Pedro II during the emperor's visit to Boston. Recognizing Bell, the emperor now greeted him warmly in anticipation of the important demonstration. Of course, the fact that Bell was personally acquainted with the Brazilian emperor was testament to more than Bell's considerable reputation as a teacher of the deaf. It spoke to another key advantage he could parlay: a wealth of connections in academia and high society.

Because Bell had failed to register for the exposition by the deadline, his exhibit was housed far less prominently than Gray's. As a result, the delegation had to hike the length of the vast hall and then climb a flight of stairs to the remote East Gallery, where Bell's devices sat on a plain wooden table. Following behind Bell and Emperor Dom Pedro II, the group, which consisted of Sir William Thomson, the Parisian instrument maker Rudolph Koenig, the astronomers James C. Watson of Ann Arbor and Henry Draper of New York, and many others—including Elisha Gray—made their way to Bell's exhibit and huddled around his table. Several of the perspiring gentlemen collapsed gratefully into chairs that Bell had placed nearby.

Bell began by explaining the theory of sympathetic vibrations that had led him to his multiple telegraph and telephone inventions. In a demonstration considerably more modest than Gray's, he illustrated his multiple-messaging telegraph by showing that Sir William and His Majesty Dom Pedro II could telegraph simultaneously over one wire. As Bell recounted in a letter to his parents:

> *I then explained the "Undulatory Theory" and offered to test the transmission of the human voice. I stated however that this was*

> *"an invention in embryo." I trusted that they would recognize firstly that the pitch of the voice was audible and secondly that there was an effect of articulation.*

As planned, Bell then headed some 500 feet down the hall, leaving Willie Hubbard with the crowd at the telephone receiver. For the demonstration, Bell used a magneto-electric transmitter he had taken to calling a membrane transmitter. Its parchment diaphragm, with a small piece of metal glued to the back side, sat immediately in front of a magnet. For the receiver, Bell used his latest model, which he called an iron-box receiver. This device consisted of a small, hollow iron cylinder; running up its center was a metal rod wrapped in a coil of insulated copper wire. The iron tube was closed at the bottom; on the top it had a lid made from a loosely attached piece of sheet metal that vibrated to transmit sound when the current came through its electromagnet.

Sir William Thomson sat attentively in a chair next to the table that held Bell's devices. As Bell had directed, Sir William pressed the lid of the iron-box receiver against his ear. Then, excitedly, he began to

THE TELEPHONE DEVICES BELL DISPLAYED AT THE CENTENNIAL EXPOSITION: THE IRON-BOX RECEIVER (*LEFT*) AND THE MEMBRANE TRANSMITTER (*RIGHT*).

repeat the words he heard coming from the little cylinder: "Do you understand what I say?" Sir William bellowed incredulously. As Willie Hubbard later recounted, Sir William jumped out of his seat, exclaiming: "Where is Mr. Bell? I must see Mr. Bell!" As Willie dutifully led him off in Bell's direction, others took turns pressing their ears up against Bell's receiver.

Next in line at the receiver was His Majesty. "I hear, I hear!" he cried out, visibly startled and amazed. Bell, in a booming Shakespearean delivery that would surely have made his grandfather proud, had cleverly begun to recite the most famous—and easily recognizable—soliloquy from *Hamlet.*

"To be or not to be!" the emperor called out, as the crowd craned closer.

Elisha Gray then made his way to the small receiver. As he later testified about the incident in a deposition for the 1879 *Dowd* case:

> *The Emperor had just been using the receiving instrument, and as he took it down from his ear and started away to the transmitting end, he said, "to be or not to be." From this I took the cue as to what was being recited at the transmitting end. I listened intently for some moments, hearing a very faint, ghostly, ringing sort of a sound; but finally, I thought I caught the words, "Aye, there's the rub." I turned to the audience, repeating these words and they cheered.*

And so, according to Elisha Gray, the first words he ever heard over what we now call a telephone offered a commentary on his situation so apt only the Bard could have written it.

Bell's performance succeeded in impressing the judges and hiding the evidence of the intellectual debt he owed to Gray. As Gray would explain much later, in 1885,

> *At the Centennial Mr. Bell exhibited a telephone which transmitted articulate speech. I witnessed the exhibition. He did not*

> *use a liquid transmitter on this occasion but used an instrument very nearly resembling the receiver of my invention. . . . I did not for a moment suspect anything wrong but supposed that in the eye of the law Bell had, as my counsel had advised me, every advantage.*

Persuaded that Bell had invented his speaking telephone independently, Gray was content to cede to Bell the rights and the glory for it. Always the gentleman, Gray had no cause to do otherwise. After all, Bell had built a working invention to transmit speech before he had. In truth, at the time, Gray was far more concerned with the fate of his multiple telegraph, which, he was happy to see, appeared vastly superior to Bell's version. To Gray, with his close ties to the telegraph industry, the speaking telephone represented an interesting development, but the multiple-messaging telegraph was an invention that satisfied an important commercial need and could reap a handsome and immediate financial return.

As the historian David Hounshell has persuasively explained, Gray, like so many others in his day, viewed the telephone as little more than "a scientific toy." He did not seem to anticipate the extent to which the judges would find the transmission of speech even more exciting than the transmission of music. And, even more important, as Hounshell contends, Gray misjudged the telephone's extraordinary long-term commercial potential, at least at first.

There is no doubt that Hounshell's interpretation is, at least to some extent, correct. The prevailing view of the telegraph industry in 1876 held that it was imperative to use existing wires more efficiently, sending ever more simultaneous messages at a time. As Gray wrote to his patent attorney around this time,

> *Bell has talked so* much *and done so little* practically . . . *I am working on an Octoplex between Philadelphia and New York—four [messages] each way simultaneously—eight [messages] at*

once. I should like to see Bell do that with his apparatus [emphasis in the original].

Judged from this perspective, given the wide availability of so-called duplex messaging by 1876, Bell's telephone represented a kind of step backward. As hard as it is to imagine today, this industry perspective stemmed largely from the conceptual stumbling block that, at the time, telegrams were sent from *telegraph offices.* Around this period, for instance, one Western Union official in a particularly expansive mood explained to the *New York Times* that eventually telegraph operators might "transmit the sound of their own voice over the wires, and talk with one another instead of telegraphing." Needless to say, the idea of telegraph operators talking to one another, however novel, was not viewed by many in the industry as a particularly commercial priority.

Nonetheless, the fact that Gray may have initially underestimated the importance of the telephone does nothing to alter the fact that his liquid transmitter design had paved the way for Bell's success. Over time, as more of the facts of the case emerged and the dazzling commercial potential of the telephone became clear, Gray realized that Bell had deceived him. Eventually, he would, in his stoic, midwestern way, become quietly outraged about the matter, as well he should have been. As Gray would put it in 1901,

Recently, and long since the oral statements were made and the letters referred to were written, and long since I gave my deposition in the Dowd case, facts came to my knowledge that convinced me I was wrong in assuming and believing Prof. Bell had fairly made his invention. . . . I supposed my discovery remained a secret at the Patent Office, as it should have done, and was not known to Mr. Bell. What I now state on the subject is made in view of information which satisfies me that Mr. Bell, having obtained my secrets, claimed my discovery as his own and by this means got credit for my invention.

Thanks largely to Bell's masterful performance at the Centennial Exposition, however, it would be many years before this realization would dawn on his unfortunate rival.

WITH THE ORDEAL of the Centennial behind him, Bell's spirits lifted markedly. Upon returning to Boston, he finished up his courses, graded his students' exams, and continued to improve his telephone models. Meanwhile, as word of Bell's accomplishment spread, people came to his small lab at 5 Exeter Place from all over to learn more about his device, including Sir William Thomson himself. Bell even gave Sir William a duplicate of his magneto-electric transmitter and iron-box receiver to bring back to England. During this period in the summer of 1876, as Watson later recalled,

> *A list of the scientists who came to the boarding house to see the telephone would read like the roster of the American Association for the Advancement of Science. My old electrical mentor, Moses G. Farmer, called one day to see the latest improvements. He told me then with tears in his eyes when he first read a description of Bell's telephone he couldn't sleep for a week, he was so mad with himself for not discovering the thing years before.*

By the fall of 1876, Bell had improved his telephone enough that he and Watson could conduct the first two-way, "long-distance" conversation over a dedicated telegraph line between the offices and factory of the Walworth Manufacturing Company. After business hours on October 9, Bell sat in an office at the firm's Boston headquarters and chatted with Watson two miles away at the company's factory across the river in Cambridgeport. The *Boston Advertiser* reprinted the two men's detailed and nearly identical notes of the conversation side by side as startling evidence of the new technology's viability.

Near the end of 1876, according to Watson's autobiography, Gardiner Hubbard quietly approached Western Union with an offer to sell the exclusive rights to the Bell telephone patents for $100,000. Given Western Union's near monopoly on existing telegraph lines, the deal offered the most obvious and swift strategy for commercializing the telephone. But Western Union president William Orton, no doubt still holding a grudge from his previous skirmishes over the Hubbard Bill several years earlier, flatly turned down the offer.

Orton soon realized his mistake, but by then it was too late. He would forever be remembered by students of business as the man who made one of the worst corporate decisions of all time: passing up the rights to the most profitable invention in history when he could have bought them for a song.

Meanwhile, Bell began to demonstrate his speaking telephone before paying audiences. At one of the first of these events, at the Lyceum Hall in Salem, Massachusetts, on February 23, 1877, an overflow audience paid fifty cents a head to see his new invention in operation. And Bell fulfilled their desires: he amazed the crowd by talking to Watson, who sat with a small group of witnesses at Bell's Exeter Street laboratory in Boston, some eighteen miles away. According to a newspaper account, the Salem audience marveled that they were able not just to hear Watson speak, but also sing, cough, and laugh.

Gertrude Hubbard protested around this time that such well-publicized events smacked too much of P. T. Barnum for her taste. But whatever emotions this stirred in his future mother-in-law, Bell's share of that evening's proceeds alone came to $149. It was the first money he had made from the telephone. He spent most of the windfall immediately—commissioning a silver brooch in the shape of a telephone to give to Mabel.

Whatever compromises Bell may have made to secure rights to the telephone, his actions won him the prize he had sought most fervently. He and Mabel were married on July 11, 1877. It was a modest wedding at the Hubbard residence. The ceremony took place in the very room where Bell had demonstrated the notion of sympathetic vibration to Hubbard less than three years before by singing into the family piano.

One of Bell's first public demonstrations of the telephone, February 23, 1877. (*Top*) Watson is at the workshop at 5 Exeter Place in Boston. (*Below*) Bell is speaking with him before an audience at Lyceum Hall in Salem, Massachusetts, some eighteen miles away.

THE NEWLYWEDS, ALEXANDER GRAHAM BELL AND MABEL HUBBARD BELL, SHORTLY BEFORE THEIR DEPARTURE ON A FIFTEEN-MONTH HONEYMOON TO EUROPE IN THE SUMMER OF 1877.

As a wedding present, Bell gave Mabel a silver cross inlaid with pearls. He also gave her all but ten shares of his stock in the newly formed Bell Telephone Company, a shockingly unconventional gesture for that era. Mabel's 30 percent stake in the firm would eventually make her an immense fortune and allow the couple to live in comfortable splendor for the rest of their lives.

CONFERENCE CALL

STANDING AT THE podium of the wood-paneled lecture hall, I readied my materials as the audience drifted in. Well into the second semester of the academic year, the time had come to present my work at a Dibner Institute seminar. By academic standards, these sessions were relatively fancy affairs, with oil portraits adorning the walls and a lunch buffet served on a sideboard. The lectures attracted a healthy number of backpack-toting graduate students, but the mainstays in the audience were historians of science at the Dibner Institute or on the faculty at MIT or Harvard.

The first Dibner seminars I had attended struck me as needlessly obfuscating and arcane. The speakers, always deeply immersed in their subject matter, made few concessions to the prospect that audience members might not be experts in their narrow specialty. But the seminars, celebrating a seemingly old-fashioned blend of collegiality and rigorous scholarship, had a quirky kind of charm that had grown on me over the course of the year.

George Smith, the acting director, seemed to especially enjoy his role as impresario. He invariably made a genial introduction that praised the speaker and lent some context to the day's topic. The esoteric subject matter varied widely, but somehow, one or two audience members always seemed to possess astoundingly detailed knowledge about the subject at hand. At a seminar on the history of technology used to explore the ocean floor, for example, the audience included an engineer who had designed some of the equipment in question; another on the origins of quantum theory led several aging physicists in the audience to offer their own firsthand reminiscences on the history being discussed.

My presentation was no exception. As seats began to fill, a distinguished-looking gentleman approached me carrying a box of beautifully crafted, working reproductions he had made of Bell's early telephone models. If I was interested, he said, I was welcome to use them as visual aids in my presentation. I accepted them gratefully as promising tokens of the kind of knowledgeable and engaged audience I could expect. Before long, the room, which only held about fifty, was overflowing with the largest crowd of the year. My title, "Did Bell Steal the Telephone?," had presumably piqued a bit more curiosity than usual.

I opened my remarks, naturally enough, where the story had begun for me. On the screen in the seminar room, I displayed a number of pages from Bell's notebook. I showed examples of his work in 1875 featuring varied arrangements of reeds, tuning forks, and electromagnets. I showed Bell's unexpected decision, on March 8, 1876, to introduce a cup of acidic water into his experiments. And I showed the fateful sketch for a liquid transmitter Bell penned in his notebook on March 9—the very eve of his successful call to Watson. When I displayed Bell's sketch side by side with the drawing Gray had previously filed in his caveat, the uncanny resemblance drew an audible gasp from the crowd.

"We remember Bell as the inventor of the telephone, but a closer look reveals an intriguing story that stands at odds with the commonly accepted tale," I said.

"Based on my research so far," I told the audience, "I can claim with a high degree of certainty that Alexander Graham Bell *did not* invent the telephone transmitter he so famously used to call to Watson.

"Rather, the evidence strongly suggests that Bell stole the design from his rival Elisha Gray."

The audience, accustomed to a diet of dense academic discourse, seemed to relish a dish of intrigue. With the heady feeling familiar to any storyteller, I could sense from the preternatural silence that I had caught the full attention of everyone present. And to this receptive crowd, I proceeded to report what I had learned about how Bell plagiarized Gray's design and how he managed to get away with it.

I told them, in other words, the same story I have recounted in these pages: a story about a driven and talented young man who became caught up in a web of family ties, business pressures, and unsettling, unexpected love.

I told them about a time, in the 1870s, when Victorian sensibilities collided with stunning new capabilities that were not yet fully understood; a horse-and-carriage era of hucksters and patent medicines that found itself contending with the power of the steam engine and the near-global reach of the telegraph. I explained that, with the telegraph industry's exponential growth, its corporate owners desperately sought—and would pay handsomely for—a multiple-messaging telegraph, a goal that captured the minds of some of the greatest inventors of the day, including Thomas Edison, Elisha Gray, and, of course, Alexander Graham Bell.

The truth is, I had a good deal of ground to cover. After stumbling upon the incriminating sketch in Bell's notebook, I had managed to piece together far more about Bell's story than I could ever have imagined at first. I had found evidence that Bell's patent was filed under highly irregular circumstances and that its original version had suspicious additions written into the margin that were never fully explained. I had found evidence clearly documenting that Bell knew about, but never fully acknowledged, the pioneering work of Philipp Reis, who had built a crude but functional telephone that antedated Bell's own model by more than a decade.

I had documented the way Bell withheld from the public—and from Gray—the truth about his path to the telephone, with actions that are all but inexplicable except as a skillful effort to cover his tracks so that Gray, distracted as he was with the invention of the multiple telegraph, wouldn't realize what Bell had done. And, of course, my research had led me to the problematic confession of an official at the U.S. Patent Office who claimed to have facilitated Bell's plagiarism and awarded him an airtight patent on an invention Bell had not, at the time, properly reduced to practice.

Reviewing all this evidence, I told the assembled group that perhaps the most troubling finding is the likelihood—suggested but not definitively proven—that Bell stole from Gray more than just the design of the liquid transmitter: that Bell's conception of varying resistance as a means to convert sound into an electrical signal only crystallized for him after he had seen Gray's work. If true, this hypothesis, suggested by the fact that the only discussion in Bell's patent of variable resistance occurs in addendums written into its margins, would significantly diminish Bell's claim to the telephone by casting doubt on a key conceptual development that is widely attributed to him.

IN AUGUST 1877, a month after their wedding, Alec and Mabel Bell left for an extended, fifteen-month trip to Europe. (She had convinced him his name was more dignified without the "k.") To fund the trip, Bell sold a Providence, Rhode Island–based entrepreneur the rights to commercialize his telephone in Britain for $5,000 in cash. Ironically, in so doing, Bell forfeited the same rights he had held out so assiduously to try to secure in his failed arrangement with George and Gordon Brown.

After marrying Mabel, Bell played virtually no further role in the technical development of the telephone. Bell's biographers tend to downplay this remarkable fact as a simple reflection of his disinterest in business; but such an explanation has never sufficiently accounted for his actions, especially those in the first several years of his marriage.

My year working alongside historians underscored for me that the reinterpretation of historical evidence can sometimes resolve outstanding questions in the historical record that have never been adequately explained. I believe this is such a case: that an understanding of Bell's likely guilt and revulsion over his underhanded role in the race to secure his telephone patent sheds new light on the extraordinary way in which he shunned any role in the newly formed company that bore his name.

Consider, for instance, the emotional language contained in a letter Bell wrote Mabel from London in September 1878, while he was on an excursion to visit a school for the deaf in Scotland:

> *Of one thing I am quite determined and that is to waste no more time and money on the telephone. . . . I am sick of the telephone and have done with it—excepting as a plaything to amuse my leisure moments. . . . There is too much of the element of speculation in patents for me. A feverish anxious life like that I have been leading since our marriage would soon change my whole nature. Already it has begun injuring me and I feel myself growing irritable, feverish and disgusted with life.*

The catalyst for this outpouring is revealing. The *Dowd* lawsuit had begun, accompanied by questions in the press about Bell's rightful claim to the telephone. In that case, Western Union had established a subsidiary (headed by Peter Dowd) which had begun offering telephone service, claiming the legal cover of a collection of telephone-related patents licensed to them by Elisha Gray, Thomas Edison, and several others. Protecting its interests, the Bell Telephone Company sought to shut down Western Union's telephone operation. Toward this end, Bell Telephone sued Dowd—and by proxy Western Union—for patent infringement. The case was thus the first to bring the telephone-related claims of Bell and Gray into conflict in court.

The legal rules of the day required all parties to make pretrial depositions, and by September 1878, while Bell was still abroad, Hubbard had begun nagging him to make one for the *Dowd* case. Bell, however, seemed nothing short of desperate to avoid testifying under oath about his role in the development of the telephone, and even avoided a meeting with Anthony Pollok in Paris that Hubbard tried to arrange.

Bell's state of mind can be seen clearly in the events that transpired later that fall. By then, the Bell Telephone Company was threatened with having to forfeit its case in *Dowd* unless it could convince Bell to make his sworn statement to bolster the company's claim to exclusive rights on the telephone. But Bell never managed to find the time despite repeated reminders. That October, Bell wrote Hubbard to say that he, Mabel, and their new baby daughter, Elsie, would be returning from Europe. In a shocking snub to the Bell Telephone Company's interests and to the Hubbards, however, Bell wrote that he and his young family would head straight to Ontario to see Bell's family, bypassing Boston altogether.

Hubbard, becoming desperate, wrote Bell's father for help. Finally, in an effort to save the company, Hubbard even dispatched Watson to head off the young family upon their steamship's arrival in Quebec. But when Bell, Mabel, Elsie, and a nursemaid arrived on November 10, Watson recounts,

> *I found Bell even more dissatisfied with the telephone business than his letters had indicated. He told me he wasn't going to have anything more to do with it, but was going to take up teaching again as soon as he could get a position.*

Eventually, the familiar coalition of Hubbard and Mabel, aided this time by Watson on the front lines, managed to persuade Bell at the last possible moment to go to Boston to make his sworn statement. Even so, Bell insisted on stopping first at his father's house in Brantford, Ontario. Watson notably recounts that he opted to accompany Bell so as not to "run the risk of losing him."

It was clearly a difficult and pivotal moment. Ultimately, in a role she had played numerous times before, Mabel convinced Bell to return to Boston. He was ill when he left Ontario with Watson and eventually wrote out his statement for the lawsuit from the Massachusetts General Hospital where he was being treated for abscesses. Mabel and Elsie finally joined Bell in Boston. But before leaving her in-laws, Mabel wrote her mother about the matter:

> *Oh, if I could only be with him as I am so worried for the whole responsibility of his going is mine; he would not have gone but that I felt so strongly about it.*

In this remarkable episode, eerily reminiscent of his reluctance over the Centennial Exposition, Bell was persuaded to defend his claim to the telephone only after dramatic appeals from his closest friends and family, and then Mabel's intervention. In the ensuing decade, Bell would have to reprise his role in court repeatedly as the Bell Telephone Company aggressively fought its competitors in literally hundreds of legal actions designed to stamp out infringers and protect its patent-backed monopoly.

Of course, aside from underwriting Bell's comfortable lifestyle, the telephone would open countless opportunities. Bell clearly loved meeting the scientific elite of his day as a respected peer, just as he relished the chance to demonstrate personally the telephone to Queen Victoria in 1878, or to accept countless honors, such as the 50,000-franc Volta Prize from the French government in 1880.

For all the glory, though, I believe that Bell remained, on some level, deeply ashamed about the circumstances that led to his patent on the telephone. His guilt can arguably be seen to have influenced many of his choices. It can help to explain why, for instance, Bell gave his financial stake in the Bell Telephone Company to Mabel and, aside from testifying on its behalf, never sought any substantive involvement in the firm.

Bell's guilt can be seen, too, as a factor in his decision to build a lavish retreat—Beinn Bhreagh—at a notably remote spot on Cape Breton Island in Canada. Mabel lends credence to the notion in the summer of 1885, the time of their first visit there, when she writes that Bell felt he needed to get "far away" from the controversy over his disputed role in the invention of the telephone. By that time, of course, the Bell Telephone Company had become wildly profitable; but legal battles continued to rage. The U.S. government was about to begin its own case to have Bell's patent annulled on the charge that it had been obtained through fraudulent means, and Wilber's incriminating affidavit had just been published. As Mabel put it, Bell felt besieged by the accusations, saying that he felt "his life has been shipwrecked."

Bell is often fondly cited as saying that he would rather be remembered as a teacher of the deaf than as the inventor of the telephone. A good example can be seen in a booklet, published by AT&T in 1947 to mark the centennial of Bell's birth, which notes that,

> *all through [Bell's] life, he maintained a deep interest in the problems of the deaf. In fact, his modesty and humanity were such that he told his family he would rather be remembered as a teacher of the deaf than as the telephone's inventor.*

Bell no doubt felt deservedly proud of his accomplishments as a teacher of the deaf. But even this sentiment, diminishing as it does his role in the birth of the telephone, can be seen in different light.

Ultimately, in considering Bell's testimony over the years, I imagine that he came to believe the versions of the story he told under oath which claimed, in essence, that he had never behaved improperly. There is no question that Bell *had* been on the right track in his telephone research; that he had made an important contribution with his theoretical understanding of the way "undulatory currents" could carry acoustic signals electrically; and that he believed in the telephone's viability when few shared his vision. Surely, those facts must have helped to assuage whatever guilt he might have felt over the purloined liquid transmitter.

Plus, Bell must certainly have taken comfort in his lifelong happiness with Mabel. The key episodes in the telephone's invention had occurred when Bell was still in his twenties. Whatever ethical compromises he may have made in his dealings with Hubbard, Wilber, Pollok, and Bailey, Bell had mostly likely made them for love. As he wrote Mabel in 1878, on the cusp of the *Dowd* case,

> *Why should it matter to the world who invented the telephone so long as the world gets the benefit of it? Why should it matter to me what the world says upon the subject so long as I have obtained the object for which I laboured and have got you my sweet sweet darling wife?*

If the secret in Bell's notebook changes our estimation of him, it also highlights the brilliant technological innovations of his competitor,

MABEL AND ALEC BELL, CIRCA 1885.

Elisha Gray. Gray's work is now largely forgotten. But his contributions to the telephone stand up under close scrutiny as seminal and creative advances that made the invention possible.

In his day, Gray was esteemed as an inventor by his peers and financially well rewarded. He was also always a proper gentleman. But he became increasingly outspoken about the telephone affair toward the end of his life as more of the details of Bell's actions came to his attention.

Even without the benefit of the smoking-gun evidence that lay hidden in Bell's notebook, Gray pieced together much of the story from Bell's own statements and from patent examiner Zenas Wilber's remarkable confession. As Gray wrote in a letter to the journal *Electrical World and Engineer* that was published a week after his death in 1901:

> *I became convinced, chiefly through Bell's own testimony in the various suits, that I had shown him* how *to construct the telephone with which he obtained the first results [emphasis added].*

Gray wrote that he had chanced to meet Wilber on the street in New York, years before the former patent examiner's final affidavit. As Gray recalled, Wilber told him:

> *Gray, you invented the telephone, and if your damned lawyers had done their duty, you would have had it. But at that time I did not know you very well, and you had never given me any cigars or asked me out to take a drink.*

Sadly, as the evidence accumulated, Gray died embittered over the way he had been wronged. A handwritten note found among his belongings after his death poignantly summed up his resentment. Gray may well have been right when he lamented:

> *The history of the telephone will never be fully written. It is partly hidden away in 20 or 30 thousand pages of testimony and partly lying*

on the hearts and consciences of a few whose lips are sealed—some in death and others by a golden clasp whose grip is even tighter.

There was still a bit of unfinished business about Bell's role in the telephone gambit. George Smith, who has experience serving as an expert witness in court, brought one vitally important practical matter to my attention on the day after my Dibner seminar. As Smith excitedly and rightly noted, the verdict in the *Dowd* case and subsequent lawsuits might well have gone against Bell had his laboratory notebook ever been subpoenaed. In Bell's day, however, as evidenced in Bell's foot-dragging in the *Dowd* case preliminaries, courts relied exclusively on pretrial statements and depositions to establish the facts of a case. The so-called rules of discovery we now take for granted as part of our legal system would have required Bell to hand over his notebooks—including the incriminating evidence they contained. But these rules were not uniformly adopted until 1938.

Changes in the legal rules of discovery could explain why the secret in Bell's notebook never factored in any of the court cases over the telephone. But what about since then? How could the incriminating sketch in Bell's notebook have been overlooked by so many of the historians who must surely have pored over his work? As it turned out, the answer to this question was simpler than it seemed, but I would not find it until my year at the Dibner Institute was coming to an end.

Spring had arrived, and the pace and excitement of the academic year seemed to quicken as the end of the semester drew near. I had been offered a chance to do some contemporary investigative work and, at least from the perspective of 1876, I welcomed the opportunity to get back to the future. The Dibner Institute kindly allowed me to continue to use the office through the summer; but with the term ending in May, my many dozens of books were due back to the library after the luxury of having them all within easy reach on my office shelves over the course of the entire academic year. I had tried to take notes and photocopy key passages but now faced the prospect of their disappearance with a sense of closure mixed with mild panic. After all, I hoped to write about Bell's story, and some of these books were exceedingly rare old works that would not be easy for me to obtain again.

One morning, as I was trying to systematically collect my research files, I came across a lingering lists of questions I had made about the Bell case, one of many over the course of the year. It contained a question I still hadn't answered about the provenance of Bell's notebook itself. When, I had asked, had anyone besides Bell gotten a chance to look at it? My plan had been to search through the references of many of the books I had collected to find the earliest citations I could that mentioned it as a source. Now, with the books due, there wasn't nearly enough time to complete such a task. Exasperated with myself, I did something we all routinely take for granted every day: I used the telephone.

Thinking more like a journalist now, I called the switchboard at the Library of Congress and asked to speak with whomever was in charge of the Alexander Graham Bell Collection. I only hoped that the archivist I spoke with would be as concerned about issues of provenance as Roland Baumann at the Oberlin Archives had been.

I was in luck. Leonard Bruno, a curator in the Library of Congress's manuscript division, answered the phone. I began by thanking him and his library for having made the Bell Collection available online. I told him how helpful the documents had been in my work. Then I asked him what I might find about exactly when Bell's laboratory notebooks had become publicly available. Bruno told me he would look into the matter.

Within the hour, Bruno called me back. According to the accession data, he said, the Bell family papers were not made available on an unrestricted basis until the family donated them to the Library of Congress in 1976. Prior to that, they had remained in the family's possession in a special room at National Geographic headquarters in Washington, D.C. Even more impressively, Bruno had been able to locate a brief 1974 memo that went to the heart of my question. Written by Bruno's predecessor as the accession was being negotiated, the note read:

> *As far as I could determine, only one historian had ever used the papers on deposit there [at National Geographic] to any great extent.*

Hard as it was for me to believe, Bruno's records indicated that Bell's notebooks had been kept almost entirely hidden from public scrutiny for a full century; and after that, they really only became widely accessible in 1999, when they were finally posted online in digital form. Even today, Bell's key laboratory notebook remains, at least to some extent, buried in plain sight as part of a vast collection of some 147,000 documents. The information Bruno gave me went far toward explaining why Bell's incriminating sketch might never yet have received wider attention: only a relatively small number of people had ever read Bell's notebooks closely before I happened to do so.

Upon reflection, it seemed clear that the lone historian mentioned in the memo was Robert Bruce, who negotiated access to the collection with the Bell family and spent a decade on the research for his 1973 biography of Bell before the collection was donated to the Library of Congress. Bruce's work offers a masterful handling of the kaleidoscopic details of Bell's long and fruitful life. But Bruce became so partisan about the issue of Bell's priority in inventing the telephone that it seems to have blinded him to the evidence before him. Convinced by scant data that Bell deserved sole credit, Bruce wrote in 1997, for instance,

> *Though independently attested records state [Bell's] basic idea in October 1874 and its vital supplement of variable resistance in May 1875, he was bedeviled by a rival [Gray] who contested his priority by claiming to have conceived both ideas in November 1875. And there still remain some, apparently believing in time travel or the occult, who suggest that Bell stole one or both ideas.*

Bruce was the first historian with access to Bell's notebooks. But he missed one of the most fascinating pieces of Bell's story.

In the end, perhaps the most important failing is not Bell's or Bruce's, but our own. Bell's notebook aside, the most striking thing about the whole case is how much was uncovered about it even in Bell's day. Gray knew many of the details himself, and he concluded that Bell had stolen his design. And, in the intervening years, various capable people have reviewed the evidence and reached a similar conclusion. I have tried to credit the work of many of them in these pages.

None of these efforts, though, has ever managed to do much to pierce the seemingly invincible myth that Bell single-handedly invented the telephone. For many years, even after Bell's death in 1922, this myth was skillfully nurtured and promoted by a monopoly whose interests it served.

I have no doubt that Elisha Gray's contribution to the invention of the telephone is important and unduly neglected. But, in my obsession to get to the bottom of the story, my aim is not to supplant one myth with another. Bell may well have stolen Gray's design for the breakthrough variable resistance liquid transmitter, yet there is little doubt that Gray, locked as he was into his shorter-term interest in the telegraph industry's sought-after multiple telegraph, would probably have been slow to commercialize the telephone even if he did pioneer it. As I had learned from my research, both Bell and Gray owed a considerable debt to the pathbreaking work of Philipp Reis in Germany. And, to name just one of many other vital contributions, without the ensuing transmitter improvements made by Emile Berliner and Thomas Edison, the commercial, long-distance telephone industry could never possibly have gotten off the ground.

Despite the unscrupulous dealings on Bell's behalf—at least some of which were almost certainly undertaken with his collusion and consent—Bell's vision and his energy still stand out as a remarkable model. The telephone brought him world renown when he was still a young man. But in all the decades that followed, Bell never rested. He built flying machines, bred sheep to study heredity, and tried to learn what he could about the language of the Mohawk tribe who lived in

Canada not far from his parents' home in Brantford, Ontario. In just one of myriad examples of his foresight, Bell even fretted about the air pollution of his day in a manner that seems particularly prescient. The soot in the air would block some of the sun's heat, Bell figured in 1917. But he reasoned that the earth, on balance, would gain some of the heat normally radiated into space, calling it "a sort of greenhouse effect."

History is messy, and delving deeper doesn't necessarily make it come much clearer. We can pin down many of the details of what happened in the past. But it is up to us what lessons we take away. Still, if I learned anything from my research into the invention of the telephone, it is that history needs to be constantly challenged and interrogated. To do anything less is to play a game of "telephone," tacitly accepting the garbled story that is whispered from one generation to the next.

ACKNOWLEDGMENTS

OUR CAPACITY TO unearth any truths about the past depends upon the extent to which we have access to the records in question. In this regard, I want to especially acknowledge and commend the remarkable, pathbreaking job the manuscript division of the U.S. Library of Congress has done by digitizing and making freely available the Alexander Graham Bell Family Papers. It is especially fitting that this collection should so demonstrate the modern marvels of telecommunications. The originals of most of the key documents cited in this book can be found in this accessible online archive, as can thousands upon thousands more. The collection immeasurably facilitated my research and sets a wonderful example for other archives around the world to emulate.

Meanwhile, anyone who uses primary source documents or artifacts in his or her work most certainly owes a debt to the painstaking and often thankless labors of many archivists, curators, and librarians. I want particularly to acknowledge several who helped me, including: David McGee, Ben Weiss, Philip Cronenwett, Anne Battis, Howard Kennett, and the rest of the staff at the Dibner Institute's Burndy Library; John Liffen, curator of communications at the Science Museum in London; Roland Baumann, Ken Grossi, and Tamara Martin at the Oberlin College Archive; Leonard Bruno, curator in the manuscript division of the Library of Congress; Jeffrey Mifflin at the MIT Archive; and Charles Sullivan and Susan Maycock at the Cambridge Historical Commission.

I am indebted to my many friends and colleagues at MIT's Dibner Institute, which is now sadly defunct after a falling out with its host university. (The Burndy Library collection, and some continuation of the Dibner's related fellowship program, now resides at the Huntington Library in San Marino, California.) The Dibner Institute provided the crucial support and home for my initial research. I especially thank George Smith and Bonnie Edwards, respectively the acting director and executive director of the Dibner Institute while I was there, for the special haven they fostered for this kind of research and their willingness in 2004 to make room at their "historians' table" for a science writer. During my year at the Dibner, I was also helped enormously by the rest of the institute staff, including Trudy Kontoff and Rita Dempsey, who smoothed many bureaucratic and logistical edges for me.

I benefited from contacts, discussions, and seminars with many of the scholars who overlapped with me at the Dibner Institute during the 2004–05 academic year, including Tom Archibald, Peter Bokulich, Alexander Brown, Claire Calcagno, Dane Daniel, Ford Doolittle, Gerard Fitzgerald, Olival Freire, Kristine Harper, Arne Hessenbruch, Giora Hon, Cesare Maffioli, Takashi Nishiyama, Prasannan Parthasarathi, Sam Schweber, Peter Shulman (no relation), Jenny Leigh Smith, Katrien Vander Straeten, Jim Voelkel, Sara Wermiel, and Chen-Pang Yeang. For special technical and moral support, I offer particularly sincere thanks to David Cahan and Conevery Valencius, both of whom went far above and beyond the call of duty to make extensive and tremendously helpful suggestions to an early draft of this manuscript. I learned a great deal from the impressive scholarship undertaken by all of these dedicated historians; but, of course, my thanks to them in no way implies their endorsement of my work. Despite their help, the interpretations and mistakes in the text, for better or worse, are mine alone.

I am particularly grateful to a number of others who also read the book in draft form and helped me to improve it in many ways, including Christopher Clarke, John Liffen, Nancy Marshall, David McGee, Dave Pantalony, and Jill Shulman. Still other friends and colleagues gave encouragement at various points along the way. Among these, I thank Dan Charles and Lewis Cohen for conversations that sparked new avenues in my research; Sarah Shulman and Tom Garrett for all their support throughout; Doug Starr at Boston University's Science Journalism Program; Marcia Bartusiak, Rob Kanigel, and Tom Levinson at MIT's Graduate Writing Program; Victor McElheny, longtime mentor and former director of MIT's Knight Science Writing Fellowship; David Talbot at MIT's *Technology Review* magazine; and Deborah Cramer, who succeeded me as the Dibner Institute's second (and, sadly, last) science writer fellow at MIT.

Marc Miller was one of my very first magazine editors and, luckily for me, he has never managed to get free from my prose since. A prize-winning historian himself, Marc has earned the dubious distinction of editing each of the five books I have written. I can't offer enough thanks for his generous and immeasurably helpful labors on my behalf any more than by calling him a dear and special friend.

My agent, Katinka Matson at Brockman, Inc., believed in this project from the start and, as usual, found a wonderful home for it. I thank Angela von der Lippe for her wise insights and editorial comments, as she, Lydia Fitzpatrick, and Sabine Eckle helped guide the book through production, and I am grateful for the sharp eye and keen judgment of Ann Adelman, who copy-edited the manuscript.

Words cannot express my gratitude to the members of my amazing family, who have once again so kindly cared for and put up with me as I wrestled a manuscript to the ground. I am a lucky father and I give my deepest thanks to Elise and Ben for everything, as well as to the rest of my extended family, especially my father, Roy Shulman, to whom this book is dedicated. As usual, I absolutely couldn't have managed without the selfless support and love of my wife and soul mate Laura Reed. She gets extra special thanks for being such an insightful and active listener and muse on all those long car rides back and forth to Boston as I worked to envision the scope and shape of this project.

In keeping with the theme of this book, I want finally to acknowledge my debt to the many researchers over the years who, after combing through the archival material about the telephone's origins, felt compelled, as I did, to try to revise the widely held misperception that Alexander Graham Bell was the telephone's sole inventor. The list includes George B. Prescott in 1878; Silvanus Thompson in 1883; John Paul Bocock in 1900; William Aitken in 1923; Lloyd Taylor in 1937; Lewis Coe in 1995; Edward Evenson, and Burton Baker, both of whose works appeared in 2000. That some of these efforts date to the earliest days of the telephone points up the long-standing roots of the controversy over its invention and, of course, also underscores just how difficult it is to correct the record adequately in the face of persistent historical myths. Nonetheless, I have learned from and taken inspiration from these researchers' efforts to ferret out the truth; each of them made contributions that have furthered the public understanding of a complex story. If, in my odyssey through this material, I have added to their labors, it is perhaps only by pointing out the significance of the information contained in Bell's notebook itself, and by weaving and culling together disparate details, many of which one or more of these researchers had already sought to bring to light.

NOTES

1: Playing Telephone

11 **5 Exeter Place:** Bell moved to two rooms (Ns. 13 and 15) at this location in mid-January 1876. See Alexander Graham Bell (AGB) to Mabel Hubbard, January 17, 1876, his first letter from the new location. Unless otherwise noted, letters come from the vast digitized collection called the Alexander Graham Bell Family Papers, Library of Congress (LOC), available online at http://memory.loc.gov/ammem/bellhtml/bellhome.html.

11 **a metal cone:** Bell's diagram and specifications are available in AGB, "Laboratory Notebook, 1875–1876" (cited hereafter as Laboratory Notebook, 1875–1876), LOC, pp. 40–41. Watson explains much later that the original liquid transmitter does not survive as it was used for parts in successive adaptations—See Thomas A. Watson, "The Birth and Babyhood of the Telephone: An Address Delivered Before the Third Annual Convention of the Telephone Pioneers of American at Chicago, October 17, 1913" (New York: American Telephone & Telegraph Co., 1936).

13 **the "undulations" created:** See, e.g., AGB, U.S. Patent 174,465, "Improvements in Telegraphy," filed Feb. 14, 1876; issued March 7, 1876. Bell claimed, among other things, a "method of producing undulations in a continuous voltaic circuit by the vibration or motion of bodies capable of inductive action."

14 **"I then shouted":** AGB, Laboratory Notebook, 1875–1876, LOC, pp. 40–41.

14 **Watson's notes:** Watson's notebook is part of the AT&T Historical Collection, NY. The page in question is printed in a full-scale color reproduction in H. M.

Boettinger, *The Telephone Book: Bell, Watson, Vail and American Life 1876–1976* (Croton-on-Hudson, NY: Riverwood Publishers, 1977), p. 67.

14 **"I feel that I have at last found":** AGB to Alexander Melville Bell, March 10, 1876.

15 **"This wire":** Reprinted in Thomas A. Watson, *Exploring Life* (New York: D. Appleton & Co., 1926), pp. 126–27.

2 : Disconnected

18 **"wizard of Menlo Park":** According to one of Edison's biographers, Matthew Josephson, the appellation was first given in an article entitled "An Afternoon with Edison," *New York Daily Graphic,* April 2, 1878. See Matthew Josephson, *Edison: A Biography* (New York: John Wiley & Sons, 1959), p. 170.

18 **born just twenty days apart:** Edison was born on February 11, 1847; Bell on March 3, 1847.

18 **just three months of formal schooling:** See Neil Baldwin, *Edison: Inventing the Century* (New York: Hyperion, 1995), p. 25.

18 **slavishly long hours:** Ibid., p. 323.

18 **punch a time clock:** See, e.g., Seth Shulman, "Unlocking the Legacies of the Edison Archives," *Technology Review* (February–March 1997).

18 **His 1,093 patents set a record:** Between 1869 and 1933, the U.S. Patent Office issued 1,093 patents to Thomas Edison, and the number still stands as the record for patents issued to a single individual. A complete accounting of Edison's patents is available online at http://edison.rutgers.edu/patents.htm.

19 **introduced Helen Keller:** See Helen Keller, *The Story of My Life* (New York: Grosset & Dunlap, 1905), pp. 18–19.

19 **launch the journal *Science*:** Bell's early support of *Science* is detailed in Robert V. Bruce, *Bell: Alexander Graham Bell and the Conquest of Solitude* (Ithaca, NY: Cornell University Press, 1973), pp. 376–78. In addition to his close personal involvement in the selection of editors, etc., Bruce estimates that Bell spent a total of some $60,000 to help keep the journal afloat in the 1880s and early 1890s.

19 **president of the National Geographic Society:** The always energetic Gardiner Hubbard played the lead in founding the National Geographic Society in 1888; in 1897, after Hubbard's death, Bell became its president—See Bruce, *Bell,* pp. 422–23.

19 **first successful airplanes:** For more on Bell's role, see Seth Shulman, *Unlocking the Sky* (New York: HarperCollins, 2002).

19 **the eugenics movement:** See, e.g., "Frontispiece: Alexander Graham Bell as Chairman of the Board of Scientific Directors of the Eugenics Record Office," *Eugenical News: Current Record of Race Hygiene,* vol. 14, no. 8 (August 1929), describing Bell as a "pioneering eugenicist." See also AGB to Charles Davenport, Eugenics Record Office, December 27, 1912. Both documents are available online at http://www.eugenicsarchive.org.

19 **women's rights:** See, e.g., AGB to Mabel Hubbard, October 5, 1875. At the end of a long, tongue-in-cheek disquisition on the subject, Bell writes his fiancée: "I suppose it will not be long before we have a woman wanting to be President of the United States! Well it is not for me to say her 'Nay'—seeing that I am a subject of Queen Victoria—a woman-sovereign—and one of the best the world has seen—so my best wishes go with her."

19 **"Mr. Bell was tall":** David Fairchild, *The World Was My Garden: Travels of a Plant Explorer* (New York: Charles Scribner's Sons, 1939), p. 290.

20 **Bern Dibner:** For an interesting biographical sketch, see I. Bernard Cohen, "Award of the 1976 Sarton Medal to Bern Dibner," *Isis,* vol. 68, no. 244 (1977), pp. 610–15.

20 **Bohr-Rosenfeld paper:** N. Bohr and L. Rosenfeld, "Field and Charge Measurements in Quantum Electrodynamics," *Physical Review,* vol. 78, no. 6 (1950), pp. 794–98.

21 **Sir Isaac Newton:** See, e.g., I. Bernard Cohen and George E. Smith, eds., *The Cambridge Companion to Newton* (Cambridge, UK: Cambridge University Press, 2002).

22 **its sensible progression:** AGB, Laboratory Notebook, 1875–1876, pp. 1–33.

23 **notes on March 8:** Ibid., p. 35.

24 **idly twisted a box:** The story is related in many places, including by Orville Wright in Marvin McFarland, ed., *The Papers of Wilbur and Orville Wright* (New York: McGraw-Hill, 1953), p. 8.

24 **Alexander Fleming:** Fleming's discovery was published in the seminal article: A. Fleming, *British Journal of Experimental Pathology,* vol. 10, no. 226 (1929). A concise, descriptive account is given in Rupert Lee, *The Eureka Moment: 100 Key Scientific Discoveries of the 20th Century* (London: British Library, 2002), p. 29.

25 **"Returned from Washington":** AGB, Laboratory Notebook, 1875–1876, p. 34.

3: On the Hook

26 **made him seem much older:** See, e.g., AGB, "Notes of Early Life," from the "Notebook of Alexander Graham Bell" *Volta Review* (1910), available online at LOC (Series: Article and Speech Files, Folder: "Autobiographical Writings," 1904–1910, undated). As Bell recalls, ever since visiting his grandfather for a year at age fifteen, people invariably thought he was older than he actually was. Bell's pupil Mabel Hubbard thought him at least ten years older than he really was—see Mabel Hubbard Diaries, January 1879, available online at LOC.

27 **an astonishing 12,000 miles of track:** *Harper's New Monthly* (February 1876), p. 465, available online at http://cdl.library.cornell.edu/moa/. *Harper's* reports on the latest figures from *Railroad Gazette,* noting that the major railroad lines in the United States added 6,202 miles of track in 1872; 3,276 miles in 1873; 1,664 miles in 1874; and 1,150 miles in 1875.

28 **William "Boss" Tweed:** For a thumbnail summary, see "Tweed, William Marcy," in *Biographical Directory of the U.S. Congress*, available online at http://bioguide.congress.gov.

28 **Jesse James:** See, e.g., Kathleen Collins, *Jesse James: Western Bank Robber* (New York: Rosen Publishing Group, 2003), p. 30.

28 **The broad avenues:** The description of Washington is derived in part from Catherine MacKenzie, *Alexander Graham Bell: The Man Who Contracted Space* (New York: Grosset & Dunlap, 1928), chap. 1.

28 **yet to be paved:** See "A Timeline of Washington, DC History," available online at http://www.h-net.org/~dclist/timeline1.html.

28 **"city of magnificent intentions":** Ibid.

29 **doubled in population:** Ibid.

29 **half-finished Washington Monument:** See George Olszewski, *A History of the Washington Monument* (Washington, DC: Department of the Interior, 1971), available online at http://www.nps.gov/archive/wamo/history/.

29 **"Mr. Pollok has the most palatial residence":** AGB to Alexander Melville Bell, February 29, 1876.

29 **Pollok's Gilded Age mansion:** Ibid.

29 **"You can hardly understand":** Ibid.

30 **"If I succeed in securing":** Ibid.

31 **working models of inventions:** In the nineteenth century, the United States was the only industrializing nation that required patent applicants to submit a model along with a description and detailed drawing of their invention. But, according to the U.S. Patent Office, "Two fires and the general chaos of the Civil War" threatened the government's collection of models, and "their enormous quantity made them an unwanted nuisance in the 1880s." Information available online at http://uspto.gov/web/offices/com/speeches/02-11.htm. See also Seth Shulman, *Owning the Future* (Boston: Houghton Mifflin, 1999), p. 6.

31 **James E. English:** See "United States Patent Laws: Main Points of the Senate Bill Amending the Present Laws," *New York Times,* February 15, 1876, p. 1.

31 **Timothy Stebins:** Timothy Stebins, U.S. Patent 181,112, "Improvement in Hydraulic Elevators," filed March 2, 1876; issued August 15, 1876.

31 **William Gates:** William Gates, U.S. Patent 174,070, "Improvement in Electric Fire Alarms," filed April 1, 1874; issued February 29, 1876.

31 **just a few dozen patent examiners:** The *Congressional Directory* (Washington, DC: Office of the Librarian of Congress, 1876), lists twenty-two "examiners" in the U.S. Patent Office (including Zenas F. Wilber), one "Examiner of Interferences," and three "Examiners-in-Chief."

31 ***tens of thousands* of patent applications:** According to U.S. government information, 15,595 patents were issued in 1876. The number annually more than doubled between 1866 and 1896. For a table and discussion, see Thomas P. Hughes, *American Genesis* (New York: Viking, 1989), pp. 14–15.

32 **Emile Berliner's 1877 patent application:** Emile Berliner, U.S. Patent 463, 569, for a "Combined Telegraph and Telephone" (microphone), filed June 1877; issued November 1891.

32 **known as a technological footnote:** See, e.g., Daniel S. Levy, "Man-made Marvels," *Time,* December 4, 2000.

32 **one of the leading electrical researchers:** See David A. Hounshell, "Bell and Gray: Contrasts in Style, Politics and Etiquette," *Proceedings of the IEEE,* vol. 64, no. 9 (September 1976), pp. 1305–14.

33 **roughly seventy patents:** See Robert Bruce, "Elisha Gray," in John Garraty and Mark Carnes, eds., *American National Biography,* Vol. 9 (Oxford: Oxford University Press, 1999), pp. 441–42.

33 **born in 1835:** Ibid. See also George B. Prescott, "Sketch of Elisha Gray," *Popular Science Monthly* (November 1878), pp. 523–28.

33 **for an improved telegraph relay:** Elisha Gray, U.S. Patent 76,748, "Improvement in Telegraph Apparatus," issued April 14, 1868.

33 **Barton & Gray:** See *American National Biography,* Vol. 9, p. 441.

34 **a one-third interest :** See David A. Hounshell, "Elisha Gray and the Telephone: On the Disadvantages of Being an Expert," *Technology and Culture* (April 1975), p. 138.

34 **"Gray was electrician":** Watson, *Exploring Life,* p. 60. "Electrician" was the contemporary term for what we would call an "electrical engineer" today.

34 **called a "caveat":** The U.S. Patent Office rules are laid out clearly in Webster Elmes, *The Executive Departments of the United States at Washington* (Washington, DC: W. H. & O. H. Morrison, 1879), chap. 28, "The Patent Office," pp. 471–87.

34 **by 1910:** See timeline in U.S. Patent and Trademark Office, *The Story of the United States Patent and Trademark Office* (Washington, DC: U.S. Government Printing Office, 1981), p. 21.

35 **Gray filed his claim:** Elisha Gray, U.S. Patent Office Caveat, "Instruments for Transmitting and Receiving Vocal Sounds Telegraphically," filed February 14, 1876.

36 **one of the largest and most lucrative monopolies:** AT&T, the direct descendant of Bell's original agreement with Hubbard and Sanders, was incorporated on March 3, 1885. A vertically integrated monopoly, AT&T would soon thereafter become the largest corporation in the world. See Irwin Lebow, *Information Highways and Byways: From the Telegraph to the 21st Century* (New York: IEEE Press, 1995), p. 41.

4 : Calling Home

39 **to have his pocketwatch cleaned:** Edwin S. Grosvenor and Morgan Wesson, *Alexander Graham Bell: The Life and Times of the Man Who Invented the Telephone* (New York: Harry N. Abrams, 1997), p. 16.

39 **born in 1847 in Edinburgh:** Bruce, *Bell*, p. 16. Bell was born on March 3, 1847, at 16 South Charlotte Street, Edinburgh.

39 **Alexander Bell, taught elocution:** MacKenzie, *Alexander Graham Bell*, p. 13.

39 **David Bell:** Ibid., p. 32.

40 **Alexander Melville Bell:** See AGB, "Notes of Early Life." See also MacKenzie, *Alexander Graham Bell*, pp. 17–20.

40 **"Visible Speech":** See Alexander Melville Bell, *Visible Speech: The Science of Universal Alphabetics* (London: Simpkin, Marshall & Co., 1867).

40 **Shaw's preface:** George Bernard Shaw, *Pygmalion* (1916; New York: Washington Square Press, 2001), preface, p. 19.

40 **just minutes away:** Bruce, *Bell*, p. 30.

42 **Bell invariably thereafter called the experience:** See, e.g., AGB, "Notes of Early Life."

42 **reading Shakespeare aloud:** Ibid.

42 **don a suit jacket:** Ibid. As Bell recalls: "The moment my father left for Edinburgh my grandfather sent for a fashionable tailor, and I soon found myself converted into a regular dude, resembling a tailor's picture plate of an Eton school-boy."

42 **"Sanskrit cerebral T":** Grosvenor and Wesson, *Alexander Graham Bell*, p. 23.

43 **"The best thing Bell did for me":** Watson, *Exploring Life*, p. 58.

43 **Shakespearean repertory theater troupe:** Ibid., pp. 250–96.

43 **a gifted musician:** See MacKenzie, *Alexander Graham Bell*, p. 22. She says that Bell, as a boy "enjoyed music, and it was his solace all his life."

43 **following Signor Bertini's model:** AGB, Address Before the Telephone Society of Washington, February 3, 1910.

43 **Eliza Symonds Bell:** AGB to Mabel Hubbard, August 1, 1876. See also Bruce, *Bell*, p. 22.

43 **a renaissance in the science of acoustics:** See, e.g., Stephan Vogel, "Sensation of Tone, Perception of Sound, and Empiricism: Helmholtz's Physiological Acoustics," in David Cahan, ed., *Hermann von Helmholtz and the Foundations of Nineteenth-Century Science* (Berkeley: University of California Press, 1993), pp. 259–87.

44 **Sir Charles Wheatstone:** See Daniel P. McVeigh, "An Early History of the Telephone 1664–1865," a project of the New York–based Institute for Learning Technologies, available online at http://www.ilt.columbia.edu/projects/bluetelephone/html/.

44 **most lucrative telegraph patents:** Ibid.

44 **on a visit to Wheatstone's laboratory:** AGB, "Notes of Early Life."

44 **Wolfgang von Kempelen:** See Bernd Pompino-Marschall, "Von Kempelen et al.: Remarks on the history of articulatory-acoustic modeling," *ZAS Papers in Linguistics*, no. 40 (2005), pp. 145–59.

44 **"I saw Sir Charles":** Ibid. See also Grosvenor and Wesson, *Alexander Graham Bell*, p. 17.

44 **their own "talking machine":** As recounted in AGB, "Making a Talking-Machine," unpublished article, undated, Miscellaneous Articles file, LOC.

45 **"The making of this talking-machine":** Ibid.

45 **"made it yell":** Ibid.

45 **"We heard someone above say":** Ibid.

45 **Weston House Academy:** AGB, autobiographical article, February 6, 1879.

46 **his first-ever professional experiment:** See AGB, "The Result of Some Experiments in Connection with 'Visible Speech' Made in Elgin in November 1865," Alexander Graham Bell Family Collection, LOC (Subject File Folder: The Deaf, Visible Speech, Notebooks, 1865).

46 **Alexander John Ellis:** Wesson and Grosvenor, *Alexander Graham Bell*, p. 30.

46 **Hermann von Helmholtz:** See Hermann von Helmholtz, *On the Sensation of Tone* (1863), 2nd English ed., trans. Alexander J. Ellis (New York: Dover, 1954).

46 **London Philological Society:** Wesson and Grosvenor, *Alexander Graham Bell*, p. 30.

46 **a "tuning fork sounder":** For a discussion of Helmholtz's device, see Bruce, *Bell,* p. 50.

47 **the completely erroneous conclusion:** See Bell's explanation in *The Bell Telephone: The Deposition of Alexander Graham Bell in the Suit Brought by the United States to Annul the Bell Patents* (Boston: American Bell Telephone Co., 1908), Int. 19, p. 12.

48 **" 'I thought that Helmholtz had done it' ":** Quoted in MacKenzie, *Alexander Graham Bell*, p. 41.

48 **Grandfather Alexander had died:** Ibid., p. 37.

49 **passed his entrance exams:** J. Symonds to AGB, July 26, 1868; see also Bruce, *Bell,* p. 57.

49 **had suffered from chronic maladies:** See, e.g., Wesson and Grosvenor, *Alexander Graham Bell*, pp. 30–31.

49 **Melville Bell's reputation:** Ibid., p. 58.

49 **On a visit to Boston:** See Bruce, *Bell,* p. 59.

49 **Harvard University president Thomas Hill:** MacKenzie, *Alexander Graham Bell*, p. 44.

49 **to Brantford, Ontario:** Bruce, *Bell,* p. 73.

5 : No Answer

51 **"If Gray had prevailed in the end":** Bruce, *Bell,* p. 168.

52 **the liquid transmitter makes its first appearance:** AGB, Laboratory Notebook, 1875–1876, p. 39.

52 **MIT professors:** In particular, Bell met with Professors Charles R. Cross, Lewis B. Monroe, and Edward Pickering—see Bruce, *Bell,* pp. 110 and 171.

52 **world's first public demonstration:** AGB, "Researches in Telephony," *Proceedings of the American Academy of Arts and Sciences*, May 10, 1876.

52 **Watson had sat in a warehouse:** Walworth Manufacturing Co., Cambridgeport, MA. See Watson, *Exploring Life,* pp. 91–95, and AGB, "The Pre-Commercial Period of the Telephone," speech delivered November 2, 1911, p. 16.

53 **A painting by W. A. Rogers:** The painting, depicting Bell's workshop at 5 Exeter Place as it stood in March 1877, is available in a photographic reproduction in the Gilbert H. Grosvenor Collection of Alexander Bell Photographs, LOC, Bell Collection, neg. no. LC-G9-Z2-4429-B-3.

54 **the Charles Williams machine shop:** For an evocative description of the shop, see Charlotte Gray, *Reluctant Genius: Alexander Graham Bell and the Passion for Invention* (New York: Arcade Publishing, 2006), p. 82.

54 **Boston Athenaeum Building:** The building on Beacon Street in Boston was designed by Edward Clarke Cabot; construction began in 1847. See http://www.bostonathenaeum.org.

56 **Perhaps the most compelling portrait:** See Watson, *Exploring Life*, passim.

56 **By using pictures:** AGB to his parents, May 6, 1874. See also Grosvenor and Wesson, *Alexander Graham Bell*, pp. 41–43.

56 **newspapers chronicled the success:** See "Report of a Committee on the New Method of Instruction for Deaf-Mutes," *Boston Daily Advertiser,* December 1871; and AGB Papers, LOC (Subject File Folder: The Deaf, Visible Speech, Misc., 1868–1919). See also AGB to his parents, June 22, 1873. Bell writes eagerly, "The lecture has at once placed me in a *new position* in Boston. It has brought me in contact with the *scientific minds* of the *city*" (emphasis in the original). See also Bruce, *Bell,* p. 76.

56 **at Boston University:** AGB to his parents, November 1, 1873; see also AGB to his parents, April 1874.

56 **Thomas Sanders soon agreed:** AGB to his parents, October 2, 1873. Bell writes his first letter after moving into the Sanders home. For more on Sanders, see also Bruce, *Bell,* pp. 127–31.

57 **"power of electricity":** Watson, *Exploring Life,* p. 52.

58 **in the same building:** See Josephson, *Edison,* p. 62.

58 **an electrical vote recorder:** Thomas A. Edison, U.S. Patent 90,646, "Improvement in Electrographic Vote-Recorder," issued June 1, 1869.

58 **"One day early in 1874":** Watson, *Exploring Life,* p. 54.

58 **he headed straight:** Ibid. See also Bruce, *Bell,* pp. 134–35.

59 **More and more unsightly wires:** See Hounshell, "Elisha Gray and the Telephone," *Technology and Culture,* p. 144.

60 **Joseph Stearns:** Joseph Stearns, U.S. Patent 126,847, "Duplex Telegraph Apparatus," issued May 14, 1872. See also Bruce, *Bell,* p. 93.

60 **"harmonic multiple telegraph":** The most complete explanation can be found in Bell's *The Multiple Telegraph* (Boston: Franklin Press, 1876). This monograph was prepared by Bell as a detailed narrative account of his path to his invention in accordance with the U.S. Patent Office's so-called Rule 53, which required a

detailed statement in an interference proceeding. See also AGB to George Brown, October 4, 1874.

60 **build an early prototype:** Watson, *Exploring Life*, p. 57.

6 : Operator Assistance

63 **paid a fateful visit:** AGB to his parents, October 20, 1874.

63 **tutoring Mabel:** See, e.g., Lilias M. Toward, *Mabel Bell: Alexander's Silent Partner* (Toronto: Methuen, 1984), pp. 19–20.

63 **scarlet fever:** For a good discussion, see Helen Elmira Waite, *Make a Joyful Sound: The Romance of Mabel Hubbard and Alexander Graham Bell* (Philadelphia: Macrae Smith Co., 1961), pp. 36–49.

63 **146 Brattle Street:** The house burned down long ago, but pictures taken in 1922 have survived in the Gilbert H. Grosvenor Collection of Alexander Graham Bell Photographs, LOC. See neg. nos. LC-G9-Z3-126545-AB and LC-G9-Z3-126550-AB.

63 **The well-heeled Sanders family:** Bruce, *Bell,* p. 98.

64 **Gardiner Greene Hubbard:** Fred DeLand, *Dumb No Longer: Romance of the Telephone* (Washington, DC: Volta Bureau, 1908), pp. 124–27.

64 **William Hubbard:** Ibid., p. 124.

64 **first president of the Clarke School:** Ibid.

64 **Gertrude McCurdy Hubbard:** Ibid., p. 125.

65 **to learn Hebrew:** See Toward, *Mabel Bell*, p. xvii.

65 **red velvet wallpaper:** Bruce, *Bell,* p. 126.

65 **Bell gave Hubbard a demonstration:** AGB to Alexander Melville Bell, Eliza Symonds Bell, and Carrie Bell, October 23, 1874.

65 **"I brought the subject":** Ibid.

66 **From earlier letters home:** See, e.g., AGB to his parents, dated only March 1874. In this letter, Bell writes: "I do not know if I told you that the gentleman who has introduced a Bill into Congress for the purchase of all the telegraph lines by the government on the English model is that father of one of my pupils. . . . Would it not be well to write to him about the telegraph scheme?"

67 **"I am tonight a happy man":** AGB to his parents, October 20, 1874.

67 **manufacture of shoes:** W. Bernard Carlson, "The Telephone as Political Instrument: Gardiner Hubbard and the Formation of the Middle Class in America, 1875–1880," in Michael Thad Allen and Gabrielle Hecht, eds., *Technologies of Power* (Cambridge, MA: MIT Press, 2001), pp. 25–55.

67 **specialized saws:** See Edwin Jenney, "Machinery for Sawing Staves," U.S. Patent 7,380, May 21, 1850.

68 **compiled a report:** See "In the Matter of the Postal-Telegraph Bill," Gardiner G. Hubbard's presentation before the U.S. House Committee on Appropriations, April 22, 1872 (Washington, DC: Government Printing Office, 1872).

68 ***Atlantic Monthly*:** Gardiner G. Hubbard, "Our Post-Office," *Atlantic Monthly* (January 1875).

68 **the Hubbard Bill:** See "In the Matter of the Postal-Telegraph Bill," Gardiner G. Hubbard before the House Committee on Appropriations, April 22, 1872.

68 **it would be run by a consortium:** Ibid.

68 **caused controversy:** For an insightful discussion, see W. Bernard Carlson, "The Telephone as Political Instrument," in Allen and Hecht, eds., *Technologies of Power*, pp. 25–55.

68 **"franking" privileges:** See Alvin F. Harlow, *Old Wires and New Waves* (New York: D. Appleton-Century, 1936), p. 336.

69 **1876 presidential election:** For a fuller account, see C. Vann Woodward, *Reunion and Reaction: The Compromise of 1877 and the End of Reconstruction* (New York: Oxford University Press, 1991).

70 **Cambridge Gas company:** DeLand, *Dumb No Longer*, p. 124.

70 **Cambridge Railroad Company:** Robert W. Lovett, "The Harvard Branch Railroad, 1849–1855," *Cambridge Historical Society Proceedings*, vol. 38 (1959–60), pp. 23–50, cited in Carlson, "The Telephone as Political Instrument," p. 35. As Carlson notes (fn 28), partly as a result of Hubbard's work to improve the transportation and utilities, the population of Cambridge nearly doubled in the 1850s, boasting some 26,000 residents in 1860.

70 **speculating on wheat:** From an undated reminiscence by Mabel Bell in the files of the Cambridge Historical Commission, as detailed in Carlson, "The Telephone as Political Instrument," p. 34.

70 **"Mr. Thomas Sanders said":** AGB to Alexander Melville Bell, Eliza Symonds Bell, and Carrie Bell, October 23, 1874.

71 **Joseph Adams:** Bell, *The Multiple Telegraph*, p. 14.

71 **the three men formed a team:** See Bruce, *Bell*, p. 129.

71 **Pollok & Bailey:** AGB to his parents, February 21, 1875. As Bell wrote of Pollok and Bailey, "They are the most eminent men connected with the Patent Office."

72 **Hubbard urged Bell:** Gardiner Hubbard to AGB, November 18, 1874. Hubbard wrote, "I called . . . on Mr. Pollok and after discussing the matter with him, became satisfied that it was very unwise to file the Caveat, as it might do you great injury."

72 **"It is a neck and neck race":** AGB to his parents, November 23, 1874.

72 **as big as an upright piano:** Watson, *Exploring Life*, p. 62.

73 **"perfect his telegraph":** Ibid., p. 63.

73 **would file his first patent:** Bell, U.S. Patent 161,739, filed March 6, 1875; issued April 6, 1875.

73 **Bell's book-length deposition:** *The Bell Telephone: The Deposition of Alexander Graham Bell in the Suit Brought by the United States to Annul the Bell Patents* (cited hereafter as *Deposition of Alexander Graham Bell*).

73 **"Most of the 149 volumes":** Bruce, *Bell,* p. 501.

74 **a most extraordinary admission:** *Deposition of Alexander Graham Bell,* Int. 102, p. 82.

74 **Bell worked out a separate agreement:** Agreement between George Brown, John Gordon Brown, and AGB, December 29, 1875, AGB Family Papers, LOC (Subject File Folder: The Telephone, Brown, George, 1875–1888).

74 **George Brown had left by ship:** South Street Seaport Museum, New York, shipping records for the *Russia,* Cunard Line, cited in A. Edward Evenson, *The Telephone Patent Conspiracy of 1876* (Jefferson, NC: McFarland & Co., 2000), p. 70.

74 **"did not hear from Mr. Brown":** *Bell Telephone Co. et al. v. Peter A. Dowd,* Circuit Court of the U.S., District of Massachusetts, filed September 12, 1878, p. 435.

75 **"Mr. Hubbard, becoming impatient":** *Deposition of Alexander Graham Bell,* Int. 102, p. 82.

75 **Hubbard was present:** MacKenzie, *Alexander Graham Bell,* p. 111.

75 **"It is understood":** Agreement between George Brown, John Gordon Brown, and AGB, December 29, 1875.

7: Clear Reception

78 **a newspaper article:** "Bell and Helmholtz Meet," *New York Daily Tribune,* Wednesday, October 4, 1893.

79 **"the danger of Whiggism":** For an interesting discussion of the implications of Whiggism on the field of chemistry, see Jan Golinski, "Chemistry," in Roy Porter, ed. *The Cambridge History of Science,* vol. 4: *Science in the Eighteenth Century* (Cambridge, UK: Cambridge University Press, 2001), pp. 375–97.

80 **Lord Rayleigh's classic:** John William Strutt, Baron Rayleigh, *The Theory of Sound* (London: Macmillan & Co., 1877–78).

80 ***Manual of Magnetism*:** Daniel Davis, Jr., et al., *Davis's Manual of Magnetism* (Boston: Daniel Davis, Jr., 1842).

80 ***Wonders of Electricity*:** J. Baile, *Wonders of Electricity* (New York: Scribner Armstrong & Co., 1872). Bell mentions his use of this source in *The Multiple Telegraph,* p. 7.

80 **"Some years hence":** Bell, *The Multiple Telegraph,* p. 7. See also Bruce, *Bell,* pp. 104–05.

81 **"The search for truth":** MacKenzie, *Alexander Graham Bell,* p. vii.

82 **"No finer influence":** Watson, *Exploring Life,* p. 57.

82 **from elocution to table manners:** Ibid., p. 58. As Watson colorfully puts it, "Up to that time, the knife had been the principal implement for eating in my family and among my acquaintances. . . ."

83 **"We accomplished little":** Ibid., p. 57.

83 **"my faith in the harmonic telegraph":** Ibid., p. 61.

83 **"never would have continued":** AGB to Mabel Hubbard Bell, September 9, 1878.

8: Person-to-Person

84 **became a frequent visitor:** See, e.g., Gertrude Hubbard to AGB, August 20, 1875, in which she writes: "Shall we see you as usual on Sunday afternoon?"

84 **generous helpings of roast beef:** Waite, *Make a Joyful Sound,* p. 85.

84 **Bell's feelings for Mabel:** For a personal account, see AGB to his parents, June 30, 1875, in which Bell writes, "It is now more than a year ago since I first began to discover that my dear pupil, Mabel Hubbard, was making her way into my heart."

85 **"did not think him exactly a gentleman":** Mabel Hubbard Diaries, January 1879.

85 **Mabel's occasional letters:** Mabel Hubbard to her mother, Vol. 78, available at the Alexander Graham Bell National Historic Site, Baddeck, Nova Scotia.

85 **"insisted on taking me to the streetcar":** Mabel Hubbard to her mother, February 3, 1874.

85 **"What do you think":** Ibid.

85 **Mr. Bell said today:** Mabel Hubbard to her mother, November 19, 1873.

86 **attended a dance party:** See Toward, *Mabel Bell,* p. 22.

87 **"with the greatest ease":** Gertrude Hubbard to Gardiner Hubbard, February 14, 1874.

87 **started a special journal:** A copy of this journal can be found as Journal by AGB and Melville James Bell, from 1867 to August 26, 1875, LOC (Series: Miscellany; Folder: Miscellaneous Writing and Copies of Correspondence, 1868–1875).

87 **"I do not know how or why":** AGB to Gertrude Hubbard, August 1, 1875 (emphasis in the original).

87 **"I value a gentle loving heart":** Ibid.

88 **a significant breakthrough:** AGB to his parents, June 30, 1875. See also Bruce, *Bell,* pp. 145–49.

88 **Bell guessed correctly:** *Deposition of Alexander Graham Bell,* Int. 68, p. 59.

89 **As Bell would later contend:** Ibid.

89 **"I am like a man in a fog":** AGB to his parents, June 30, 1875.

89 **"I hadn't been in love":** Watson, *Exploring Life,* p. 108.

89 **"Pardon me for the liberty":** AGB to Gertrude Hubbard, June 24, 1875.

90 **prompted an immediate meeting:** AGB, Journal entry, June 25, 1875.

90 **"Called on Mr. Hubbard":** AGB, Journal entry, June 27, 1875.

91 **On a lovely June evening:** AGB to Mabel Hubbard, August 8, 1875.

91 **Bell wrote later:** AGB to Mabel Hubbard, August 8, 1875.

91 **Bell went again to Mrs. Hubbard:** AGB, Journal entry, August 4, 1875.

91 **she had received a letter from Mabel:** AGB, Journal entry, August 3, 1875.

91 **"I think I am old enough":** Mabel Hubbard to her mother, August 2, 1875.

92 **"The letter which was read":** AGB to Mr. and Mrs. Gardiner Hubbard, August 5, 1875.

92 **"regret this new burst of passion":** Gardiner Hubbard to AGB, August 6, 1875.

92 **Ocean House Inn:** AGB, Journal entry, August 7, 1875. Bell writes that the island was made impassable by a road washed out by a torrent of water some "forty feet wide and about three and a half inches deep."

92 **"You did not know, Mabel":** AGB to Mabel Hubbard, August 8, 1875.

93 **Cousin Mary informed Bell:** AGB, Journal entry, August 9, 1875.

93 **In the greenhouse:** AGB, Journal entry, August 26, 1875. Recounting this meeting, Bell calls it "The happiest day of my life."

93 **"Shall not record any more here":** AGB, Journal entry, August 26, 1875.

94 **"I have been sorry to see":** Gardiner Hubbard to AGB, October 29, 1875.

94 **"You are Mabel's father":** AGB to Gardiner Hubbard, November 23, 1875.

94 **wanted to marry Mabel:** Gardiner Hubbard to AGB, October 29, 1875.

94 **"I shall certainly not relinquish":** AGB to Gardiner Hubbard, November 23, 1875.

95 **to become formally engaged:** AGB to his parents, November 25, 1875.

95 **"I know how young":** AGB to Gertrude Hubbard, August 1, 1875.

95 **"Professor Bell had a special trouble":** Watson, *Exploring Life*, pp. 108–09.

9 : Interference

97 **includes 147,000 documents:** Grosvenor and Wesson, *Alexander Graham Bell*, p. 8.

97 ***The Health of the Country*:** Conevery Bolton Valencius, *The Health of the Country: How American Settlers Understood Themselves and Their Land* (New York: Basic Books, 2002). The book won the George Perkins Marsh Prize for the best environmental history of 2002.

98 **She offered an example:** Conevery Bolton Valencius (with Peter J. Kastor), "Sacagawea's Cold," presentation at the Conference on Health and Medicine in the Era of Lewis and Clark, College of Physicians, Philadelphia, November 2004. A related article is forthcoming from the *Bulletin of the History of Medicine* (Summer 2008).

100 **The caveat filed by Elisha Gray:** Elisha Gray, U.S. Patent Office Caveat, filed February 14, 1876.

100 **U.S. Patent No. 174,465:** AGB, U.S. Patent No. 174,465, "Improvements in Telegraphy," filed February 14, 1876; issued March 7, 1876.

103 **a full-blown congressional investigation:** At the direction of the U.S. Congress, the Department of the Interior undertook an investigation into the circumstances surrounding Bell's patent. See *Report of the Department of the Interior*, December 22, 1885. This report, issued by George Jenks, Assistant Secretary of the Interior, is included in M. B. Philipp, *In the United States Patent Office. In the matter of the petition on behalf of Elisha Gray to re-open the interferences between A.G.*

Bell, Elisha Gray and others, before the Commissioner of Patents. Brief for petitioner. M. B. Philipp, counsel for Gray and the Gray National Telephone Co. (New York: B. Quick Print Co., 1888), p. 67.

103 **"dropped off at the U.S. Patent Office":** Gray, *Reluctant Genius*, p. 121.

104 **"The caveat was prepared deliberately":** See letter from Elisha Gray, published in *Electrical World and Engineer,* February 2, 1901, p. 199.

104 **A. Edward Evenson:** Evenson, *The Telephone Patent Conspiracy of 1876*, pp. 73–83.

105 **We know this from correspondence:** Part of the so-called file wrapper of AGB U.S. Patent 174,465, now at the U.S. National Archives. Text available in Evenson, ibid., pp. 78–80; facsimiles reprinted in Burton H. Baker, *The Gray Matter: The Forgotten Story of the Telephone* (St. Joseph, MI: Telepress, 2000), pp. A25–A45.

105 **the U.S. patent system:** See U.S. Patent Code, 1839.

106 **Wilber mailed notice:** Patent Office memo to Pollok & Bailey, February 19, 1876, file wrapper, U.S. Patent 174,465.

106 **immediately responded:** Pollok & Bailey to the Commissioner of Patents, undated, file wrapper, U.S. Patent 174,465.

106 **the *Essex* case:** U.S. Patent Office, "Commissioner's Decision of February 3, 1876," *Official Gazette,* March 14, 1876, p. 497. Reprinted in Baker, *The Gray Matter*, p. A64.

106 **"There is nothing in the records":** Ibid.

107 **"We respectfully request":** Pollok & Bailey to the Commissioner of Patents, undated, file wrapper, U.S. Patent 174,465.

107 **"The regular practice":** Ibid.

109 **in an unusual governmental report:** *Report of the Department of the Interior,* p. 67.

109 **"If passing through a forest":** Ibid., p. 69.

10: Caller I.D.

110 ***The Speaking Telegraph*:** George B. Prescott, *The Speaking Telegraph, Talking Phonograph and Other Novelties* (New York: D. Appleton, 1879).

111 ***Who Invented the Telephone?*:** William Aitken, *Who Invented the Telephone?* (London: Blackie & Son, 1939).

111 ***The Telephone and Its Several Inventors*:** Lewis Coe, *The Telephone and Its Several Inventors* (Jefferson, NC: McFarland & Co., 1995).

111 **Charles Grafton Page:** See, e.g., Aitken, *Who Invented the Telephone?*, p. 7. For more on Page's life and times, see Robert C. Post, *Physics, Patents, and Politics: The Washington Career of Charles Grafton Page, 1838–1868*, Ph.D. dissertation, University of California, Los Angeles, 1973.

111 **in Bell's first public talk:** Bell, "Researches in Telephony," *Proceedings of the Amer-*

ican Academy of Arts and Sciences, May 10, 1876, p. 1. Bell's notes cite the works of many researchers, including Page. See fn p. 1 citing C. G. Page, "The Production of Galvanic Music," *Silliman's Journal* (1837), vol. 32, pp. 396, 354; vol. 33, p. 118.

111 **"carefully studied by Marrian":** Bell, "Researches in Telephony," pp. 1–2.

112 **Bell disavowed much knowledge:** *Deposition of Alexander Graham Bell,* Cross-Int. 384, p. 256. Bell notes: "I may say that previous to the issuance of my patent of March 7, 1876, I was very ignorant of the literature relating to the production of sound by electrical means."

112 **an 1854 article:** See Charles Bourseul, "Transmission électrique de la parole," *L'Illustration Journal Universel,* August 26, 1854.

112 **"I have":** Ibid., p. 139.

113 **two distinct camps:** The History of Science Society, with its journal *ISIS,* was founded in the 1920s. The Society for the History of Technology, with its journal *Technology and Culture,* was begun in 1958. Each group has its own separate memberships, conferences, and awards.

114 **focusing on Rudolph Koenig:** See David Pantalony, "Rudolph Koenig's Workshop of Sound: Instruments, Theories, and the Debate Over Combination Tones," *Annals of Science,* vol. 62, no.1 (January 2005), pp. 57–82.

115 **very early "needle telegraphs":** For more on Cooke and Wheatstone's needle telegraphs, see B. Bowers, "Inventors of the Telegraph," *Proceedings of the IEEE,* vol. 90, no. 3 (March 2002), pp. 436–39. Photographs from the Science Museum's collection are available at http://www.ingenious.org.uk.

116 **the first transatlantic telegraph cable:** For a detailed scholarly account, see Bern Dibner, *The Atlantic Cable* (Norwalk, CT: Burndy Library, 1959).

117 **the "Osborne telephone":** For a picture and description, see "Ingenious," exhibit, Science Museum, London, online at: http://www.ingenious.org.uk.

117 **Philipp Reis's story:** See Silvanus P. Thompson, *Philipp Reis: Inventor of the Telephone* (London: E. & F. N. Spon, 1883). See also John Liffen, "Precursors of the Telephone," unpublished talk given at the Royal Museum of Scotland, Edinburgh, November 18, 2003; copy courtesy of the author.

118 **by 1861, he had a telephone prototype:** For a timeline of Reis's developments, see Thompson, *Philipp Reis,* pp. 11–12.

118 **"went up into the room":** Testimonial of Heinrich Friedrich Peter, a music teacher at the Garnier Institute where Reis worked, quoted in ibid., p. 127.

119 **"I was present at the Assembly":** Testimonial of Georg Quincke, professor of physics at the University of Heidelberg, quoted in ibid., p. 112.

120 **Reis demonstrated the 1863 model:** See Basilio Catania, "The Telephon of Philipp Reis," *Antenna,* vol. 17, no. 1 (October 2004), pp. 3–8. Available online at http://www.esanet.it/chez_basilio/reis.htm.

120 **Stephen Yeates:** See, e.g., McVeigh, "An Early History of the Telephone 1664–1865."

121 **Bell's first public speech:** Bell, "Researches in Telephony," *Proceedings of the American Academy of Arts and Sciences,* May 10, 1876.

121 **Wilhelm von Legat:** Wilhelm Von Legat, "On the Reproduction of Sounds by means of Galvanic Current," *Zeitschrift des Deutsche-Oesterreichischen Telegraphen Vereins* [Journal of the Austro-German Telegraph Union], vol. 9 (1862), p. 125. *Dingler's Polytechnishces Journal*, vol. 169 (1863), p. 23, reproduced the article. An English translation appears in the "Deposition of Antonio Meucci," part III, p. 29, copy in the New York Public Library.

121 **Robert Ferguson:** See Robert M. Ferguson, *Electricity* (London and Edinburgh: William & Robert Chambers, 1867), pp. 257–58.

121 **Charles Cross:** *Deposition of Alexander Graham Bell,* Cross-Int. 327, p. 221.

122 **a well-known Edinburgh shop:** The Reis telephone was demonstrated by Mr. Shearer, dealer of "Messrs. Kemp & Co." of Edinburgh. See "The Telephon of Philipp Reis," p. 8, notes 23 and 24.

122 **a firsthand demonstration:** *Deposition of Alexander Graham Bell,* Int. 54, p. 50. See also AGB to his parents, March 18, 1875.

122 **"Before March 7, 1876":** *Deposition of Alexander Graham Bell,* Cross-Int. 843, p. 428.

123 **"Any sound will be reproduced":** Letter from Philipp Reis to William Ladd, July 13, 1863, courtesy of the Science Museum, London, inv. 1953–118.

123 **"I take the ground that":** *Deposition of Alexander Graham Bell,* Cross-Int. 854, p. 432.

123 **a demonstration in a U.S. courtroom:** See Baker, *The Gray Matter*, pp. 42–43. The issue of Reis came up in many courtrooms; but a key failure of the demonstration took place in the so-called second *Dolbear* case—Dolbear II, 17 Fed. Rep. 604, U.S. Circuit Court, District of Massachusetts, filed October 10, 1881.

124 **Silvanus Thompson argues:** See Thompson, *Philipp Reis,* pp. 165–79.

124 **British Post Office:** As detailed in Aitken, *Who Invented the Telephone?*, citing "The First Telephone," *Post Office Electrical Engineers' Journal,* vol. 25 (July 1932), pp. 116–17.

125 **a detailed series of experiments:** Letter from L. C. Pocock, Standard Telephones & Cables, Ltd., to W. T. O'Dea, assistant curator, Science Museum, October 4, 1946, copy courtesy of the Science Museum, London.

125 **STC was negotiating:** For a full review of the incident, see Liffen, "Precursors of the Telephone," unpublished talk. See also Liam McDougall, "Official: Bell didn't invent the telephone; 'top secret' file reveals that businessmen suppressed the identity of the telephone's real inventor," *Sunday Herald* (Glasgow), November 23, 2003.

11: Tapping the Phone

127 **Furthermore, by 1874:** Hounshell, "Bell and Gray," *Proceedings of the IEEE,* p. 1308.

127 **a forty-three-page booklet:** Transcript of "Complimentary Reception and Ban-

quet to Elisha Gray, Ph.D., Inventor of the Telephone," at Highland Park (Chicago: McRoy Clay Works, 1904), November 15, 1878.

127 **"The citizens of Highland Park":** Quoted in ibid., citing *The Interior,* a weekly Presbyterian newspaper in Chicago, p. 7.

128 **"If the press and the public":** Gray reception booklet, p. 10. Bingham's remarks were part of the second toast, entitled "The Telephone in Its Origin."

128 **astonished parishioners became:** Ibid.

128 **George Prescott:** Ibid., p. 38.

129 **"in coming up with":** Bruce, *Bell,* p. 170.

130 **plenty of evidence:** David A. Hounshell, "Two Paths to the Telephone," *Scientific American,* vol. 17, no. 3 (January 1981), p. 161. As Hounshell puts it, "By this time [1875] Gray and Bell were playing cat and mouse with each other. Each suspected that the other was spying on him. . . ."

130 **According to Gray's own account:** See "Deposition of Elisha Gray," in *Speaking Telephone Interferences, The Case for E. Gray* (Washington, DC, 1880), pp. 41–42.

130 **"electrotherapy" machines:** For a wealth of documents and photographs of electrotherapy machines, see the online electrotherapy museum at http://www.electrotherapymuseum.com.

131 **Gray showed the device:** Bruce, *Bell,* p. 116.

131 **widely reported in newspapers:** Ibid., p. 118.

132 **Gray patented his version:** See Elisha Gray, U.S. Patent 165,728, "Improvement in Transmitters for Electro-Harmonic Telegraphs," July 20, 1875.

132 **two boys in Milwaukee:** For the most detailed account given by Gray, see Elisha Gray, *Nature's Miracles,* vol. 1 (New York: Eaton & Mains, 1900), p. 141: "I noticed two boys with fruit-cans in their hands having a thread attached to the center of the bottom of each can and stretched across the street . . . my interest was immediately aroused. I took the can out of one of the boy's hands . . . putting my ear to the mouth of it I could hear the voice of the boy across the street. I conversed with him a moment, then noticed how the cord was connected at the bottom of the two cans, when, suddenly, the problem of electrical speech-transmission was solved in my mind."

132 **"water rheostat":** "Deposition of Elisha Gray," pp. 48–49. See also Hounshell, "Elisha Gray and the Telephone," *Technology and Culture,* p. 153.

133 **"One of Gray's staunchest supporters":** Coe, *The Telephone and Its Several Inventors,* p. 71.

133 **Taylor's one published article:** Lloyd W. Taylor, "The Untold Story of the Telephone," *American Physics Teacher* (December 1937), pp. 243–51, reprinted in Coe, *The Telephone and Its Several Inventors,* Appendix 9, p. 206.

133 **"was the first embodiment":** Lloyd W. Taylor, "The Untold Story of the Telephone," unpublished MS, Oberlin College Archives, Oberlin, OH, chap. 2.

134 **"in a confidential document":** Taylor, "The Untold Story of the Telephone," *American Physics Teacher,* p. 246.

134 **a book-length manuscript:** Coe, *The Telephone and Its Several Inventors,* p. 73.

134 **fortresslike library:** The archives are on the top floor of the Seeley G. Mudd Center, the central library facility for Oberlin College.

134 **Oberlin's extraordinary history:** See Marlene D. Merril, "Daughters of America Rejoice: The Oberlin Experiment," *Timeline: A Publication of the Ohio Historical Society* (October–November 1987), pp. 13–21.

136 **"There is no suggestion":** Taylor, "The Untold Story of the Telephone," *American Physics Teacher,* p. 247.

136 **"made and publicly used":** Ibid., p. 245.

136 **Bell had priority:** See, e.g., Bruce, Foreword, in Wesson and Grosvenor, *Alexander Graham Bell,* p. 6.

137 **"Gray had made and exhibited":** Taylor, "The Untold Story of the Telephone," *American Physics Teacher,* p. 251.

137 **"Gray's loss of credit":** Ibid.

12: Bad Connection

138 **Taylor contacted Gray's descendants:** See Lloyd W. Taylor, correspondence, 1921–1948, Untold Story of the Telephone, Folders 1–4, Lloyd W. and Esther B. Taylor Papers, courtesy of the Oberlin College Archives.

138 **the editor of the *Encyclopaedia Britannica*:** See correspondence between Lloyd W. Taylor and Walter Tust, editor of the *Encyclopaedia Britannica,* August–October 1945, Lloyd W. and Esther B. Taylor Papers, courtesy of the Oberlin College Archives.

138 **a long-forgotten trove:** See Lloyd W. Taylor, correspondence, 1921–1948, Oberlin College Archives.

139 **a revealing letter:** Elisha Gray to AGB, March 2, 1877.

139 **the *Chicago Tribune*:** "Personal column," *Chicago Tribune,* February 16, 1877. The article stated, in part, "The real inventor of the telephone—Mr. Elisha Gray, of Chicago . . . concerns himself not at all about the spurious claims of Professor Bell. . . ."

139 **"I do not know the nature":** AGB to Elisha Gray, March 2, 1877.

140 **"Zenas Fisk Wilber":** Zenas Fisk Wilber affidavit, October 21, 1885; Zenas Fisk Wilber affidavit, April 8, 1886, Thomas W. Soran, Notary Public, in Lloyd W. and Esther B. Taylor Papers, courtesy of the Oberlin College Archives.

141 **"I am convinced":** Wilber affidavit, October 21, 1885.

141 **in the same regiment:** Ibid.

141 **Columbian College Law Department:** George Washington University Law School, "Marcellus Bailey and the Telephone," available online at http://www.law.gwu.edu.

142 **"Professor Bell called upon me":** Wilber affidavit, October 21, 1885.

144 **Wilber's affidavit is extremely problematic:** See, e.g., Affidavit of John F. Guy,

September 18, 1885, Elisha Gray Collection, National Museum of American History, reprinted in Evenson, *The Telephone Patent Conspiracy*, p. 175. See also Bruce, *Bell*, p. 278. Without citing any particular evidence, Bruce dismisses Wilber's October affidavit this way: ". . . Zenas Wilber (probably liquored up or bribed, or both, by agents of the Globe Telephone Company) made affidavits that he had allowed Bell to examine Gray's caveat in full."

144 **five separate affidavits:** As reviewed in Taylor, unpublished manuscript, chap. 11.

144 **Lloyd Taylor analyzed:** For transcriptions of Zenas F. Wilber's affidavits, October 21, 1885, and April 8, 1886, see Taylor's unpublished manuscript, Appendix III.

144 **One statement by Wilber:** Wilber affidavit, October 21, 1885.

145 **"my faculties were not":** Wilber affidavit, April 8, 1886.

145 **"Extract from a letter":** As recounted in Taylor, unpublished manuscript, Appendix III, p. 2.

146 **"I have thus concluded":** Wilber affidavit, April 8, 1886.

146 **"In conclusion":** Ibid.

13: On the Line

147 **published in *The Washington Post*:** "Affidavit of Alexander Graham Bell in reply to that of Zenas Fisk Wilber," *Washington Post*, May 25, 1886.

147 **Bell's sworn retort:** Ibid.

148 **"I knew that some interference":** Report from AGB and Mabel Hubbard Bell to Gilbert Grosvenor, undated, available in the Alexander Graham Bell Family Papers, LOC.

149 **the *Dowd* case:** *Bell Telephone Co. et al. v. Peter A. Dowd*, Circuit Court of the U.S., District of Massachusetts, filed September 12, 1878.

149 **"As I knew nothing":** *Deposition of Alexander Graham Bell*, Int. 266, pp. 194–95.

150 ***is written into the margin***: See file copy of AGB Patent Application filed February 14, 1876, LOC, reprinted in Coe, *The Telephone and Its Several Inventors*, p. 6, and Baker, *The Gray Matter*, p. A76.

151 **"Strange, isn't it":** John E. Kingsbury, *The Telephone and Telephone Exchanges: Their Invention and Development* (London: Longmans, Green & Co., 1915), p. 213.

151 **"without the leading character":** Aitken, *Who Invented the Telephone?*, p. 100.

151 **Bell spent the evening of January 12, 1876:** AGB to Gardiner Hubbard, January 13, 1876.

153 **"that burglars did not enter":** Ibid.

153 **"Almost at the last moment":** *Deposition of Alexander Graham Bell*, Int. 103, p. 86.

153 **The following day:** Bruce, *Bell*, p. 165.

154 **"I have so much copy work":** AGB to Mabel Hubbard, January 19, 1876.

154 **Bell, Hubbard, and Pollok met with Brown:** Bruce, *Bell*, p. 165.

154 **George Brown's copy:** George Brown's copy of AGB's Patent Application of

1876 resurfaced as a key issue in telephone patent litigation that went to the U.S. Supreme Court. It is available at U.S. Reports 126US 88, October term, 1887. It is also reprinted in full in Evenson, *The Telephone Patent Conspiracy*, pp. 245–52, and in Baker, *The Gray Matter*, pp. A60–A63.

155 **Bell and his legal team argued:** Evenson, *The Telephone Patent Conspiracy*, p.180.

155 **"I sailed for Liverpool":** George Brown to AGB, November 12, 1878.

155 **Brown never did succeed:** See John Gordon Brown to AGB, February 27, 1876, in which he breaks the news that his brother had learned from "thoroughly competent parties" in Britain that Bell's patent application would not be viable there. See also AGB to his parents, March 10, 1876. As Bell explained: "George Brown has thrown up telegraphy as it cannot be made a commercial success in England—telegraphy being there a government concern." MacKenzie, *Alexander Graham Bell*, p. 111, notes that "Bell never forgot and never forgave" Brown for failing to pursue his patent in Britain.

156 **"Whatever the reason":** Bruce, *Bell*, p. 164.

156 **swore to before a notary public:** *Deposition of Alexander Graham Bell*, Int. 102, p. 82. As Bell states, "The American application was sworn to in Boston, on the 20th of January, 1876 and was sent to Washington and placed in the hands of my solicitors there . . ."

156 **Baker went to great lengths:** See Baker, *The Gray Matter*, pp. 117–22.

157 **"It is my firm conclusion":** Ibid., p. 132.

158 **"I have read somewhere":** AGB to Gardiner Hubbard, May 4, 1875.

158 **so-called spark arrester:** *Deposition of Alexander Graham Bell*, Int. 103, pp. 83–88.

159 **"For instance, *let mercury or some other liquid*":** AGB, U.S. Patent 174,465.

159 **"This application of the spark-arrester principle":** *Deposition of Alexander Graham Bell*, Int. 103, p. 87.

160 **no drawing or model would be necessary:** *Deposition of Alexander Graham Bell*, Cross-Int. 410, p. 265. As Bell notes in his testimony, Zenas Wilber had made the following notation on the file wrapper of his application: "The dwg. And Specn in this case are sufficient for the examination."

161 **"How did you come":** *Deposition of Alexander Graham Bell*, Int. 266, p. 195.

162 **"I do not know what it was":** Ibid.

162 **"On this gossamer thread":** Taylor, unpublished manuscript, chap. 7.

14: Call Waiting

164 **"all the Speaking Telephones . . . Mr. Gray's":** Prescott, *The Speaking Telegraph, Talking Phonograph and Other Novelties*, p. 34.

164 **the terms of the settlement:** According to Evenson, *The Telephone Patent Conspiracy*, p. 198.

165 **"all the Speaking Telephones . . . Mr. Bell's":** George B. Prescott, *Bell's Electric Speaking Telephone: Its Invention, Construction, Application, Modification and History* (New York: D. Appleton, 1884), p. 34. This version is, with several noteworthy deletions and changes, the same as his earlier work, *The Speaking Telegraph.* (Much more widely distributed than its predecessor, it was most recently available in an edition by the Arno Press, formerly a subsidiary of the New York Times Co., 1972.)

165 **"From the reading of the text":** Prescott, *The Speaking Telegraph,* p. 73, note 1.

165 **even the title of his book:** Prescott, *Bell's Electric Speaking Telephone.*

166 **"struck him a smart blow":** See Isaac d'Israeli, *Curiosities of Literature* (Paris: 1835), p. 24. For more on the apple myth, see also James Gleick, *Isaac Newton* (New York: Pantheon Books, 2003), pp. 54–57. Gleick calls it "the single most enduring legend in the annals of scientific discovery."

166 **"The apple myth":** George Smith, "Myth versus Reality in the History of Science," unpublished proposal, February 2005, quoted courtesy of the author.

167 **Irving Fang's textbook:** Irving Fang, *A History of Mass Communications: Six Information Revolutions* (Burlington, MA: Focal Press, 1997), p. 84.

168 ***The Nobel Book of Answers*:** Gerd Binning, "How Does the Telephone Work?," in Bettina Steikel, ed., *The Nobel Book of Answers: The Dalai Lama, Mikhail Gorbachev, Shimon Peres, and Other Nobel Prize Winners Answer Some of Life's Most Intriguing Questions for Young People* (Philadelphia: Chemical Heritage Foundation, 2003), p. 121.

168 **"*On 7 March 1876*":** Ian McNeil, ed., *The Encyclopaedia of the History of Technology* (London: Routledge, 1996), p. 719.

169 ***Famous Americans*:** *Famous Americans: 22 Short Plays for the Classroom* (New York: Scholastic Books, 1995), p. 93.

169 **Herbert Casson:** Casson, *The History of the Telephone* (Chicago: A. C. McClurg & Co., 1910), p. 10.

170 ***Understanding Telephone Electronics*:** Joseph Carr, Steve Winder, and Stephen Bigelow, *Understanding Telephone Electronics* (Woburn, MA: Newnes, 2001), p. 1.

170 **"A pile of tools":** Victoria Sherrow and Elaine Verstraete, *Alexander Graham Bell (On My Own Biographies)* (Minneapolis: Carolrhoda Books, 2001), p. 43.

171 **Bell made no mention of his first:** Baker, *The Gray Matter*, p. 62, citing congressional hearings, 1886.

172 **Bell does touch upon the story:** Taylor, unpublished manuscript, chap. 2.

172 **in August 1882:** Ibid.

172 **"Watson dashed down the hall":** MacKenzie, *Alexander Graham Bell,* p. 115.

173 **Bell's first public speech about the telephone:** Bell, "Researches in Telephony," *Proceedings of the American Academy of Arts and Sciences*, May 10, 1876.

173 **Scholars normally describe:** See, e.g., Karl L. Wildes and Nilo A. Lindren, *A Century of Electrical Engineering and Computer Science at MIT, 1882–1982* (Cambridge, MA: MIT Press, 1985), p. 25.

174 **a far more primitive apparatus:** AGB to his parents, May 12, 1876.

174 **"The meeting at the Academy":** Ibid.

174 **"I do not know that I can recall them":** *Deposition of Alexander Graham Bell,* Int. 118, p. 85.

175 **Instead, he switched his focus:** AGB, Laboratory Notebook, 1875–1876, pp. 81–83.

175 **a "magneto-electric" transmitter:** Bruce, *Bell,* p. 185.

176 **Bernard Finn:** See Bernard Finn, "Alexander Graham Bell's Experiments with the Variable-Resistance Transmitter," *Smithsonian Journal of History,* vol. 1, no. 4 (1966), pp. 1–16.

176 **Bell notes that it is difficult to hear:** AGB, Laboratory Notebook, 1875–1876, pp. 12–13.

176 **including most notably Thomas Edison:** See Thomas A. Edison, "Improvement in Speaking-Telegraphs," U.S. Patent 203,015, filed August 28, 1877; issued April 30, 1878. It is one of the great ironies of the history of the telephone that Bell's rival, Thomas Edison, with his invention of the carbon-button transmitter, finally perfected a commercially viable telephone design. This fact would play a large role in the final settlement of the *Dowd* suit between Bell Telephone and Western Union in November 1879.

176 **"Upon the variable-resistance transmitter":** Bruce, *Bell,* p. 185.

177 **MIT Archives:** Minutes of the May 25, 1876, meeting at MIT, "The 197th meeting of the Society of Arts," available courtesy of the MIT Archives, Cambridge, MA.

177 **the *Boston Transcript*:** *Boston Transcript,* May 31, 1876, courtesy of the MIT Archives, Cambridge, MA.

177 **"at last found the solution":** AGB to Alexander Melville Bell, March 10, 1876.

15: Party Line

179 **Grand Villa Hotel:** AGB to Mabel Hubbard, June 21, 1876.

180 **More than 30,000 exhibitors:** For a concise overview of the exhibit, see, e.g., "Progress Made Visible: The Centennial Exposition, Philadelphia, 1876," Special Collections Department, University of Delaware Library. Available online at http://www.lib.udel.edu/ud/spec/exhibits/fairs/cent.htm.

181 **"I really wish you could be here":** AGB to Mabel Hubbard, June 21, 1876.

182 **Some thirty-seven nations:** An in-depth review is available in Robert C. Post, ed., *1876: A Centennial Exhibition* (Washington, DC: Smithsonian Institution, 1976).

182 **particularly caught Bell's eye:** Bell recounts this and other exhibits in his letter to Mabel Hubbard, June 21, 1876.

182 **Frédéric Bartholdi's dramatic:** See "Colossal hand and torch 'Liberty,' " photograph in the collection of LOC, digital file reproduction no. LC-DIG-ppmsca-

02957. The official history of the Statue of Liberty is available online from the U.S. National Park Service at http://www.nps.gov/stli/.

182 **the world's first steam-driven monorail:** See John Allwood, *The Great Exhibitions* (London: Studio Vista, 1977), p. 57.

182 **newfangled elevator:** Ibid.

183 **Corliss Steam Engine:** See Post, ed., *1876,* p. 31.

184 **"loftily in the center":** W. D. Howells, "A Sennight of the Centennial," *Atlantic Monthly,* vol. 38, no. 225 (July 1876), p. 96. Available online in the "Making of America" collection at Cornell University Library, http://cdl.library.cornell.edu.

184 **Edison brought his newly designed:** U.S. Centennial Commission, *International Exhibition 1876: Official Catalogue* (Philadelphia: John R. Nagle & Co., 1876), p. 331.

184 **Rudolph Koenig:** U.S. Centennial Commission, *International Exhibition, 1876: Reports and Awards.* Vol. VII, ed. Francis A. Walker (Washington, DC: Government Printing Office, 1880), p. 12.

184 **Bausch & Lomb:** Ibid.

184 **brand-new material called asbestos:** *International Exhibition 1876: Official Catalogue,* p. 104.

184 **new tomato condiment:** According to the official history of the H. J. Heinz Co., ketchup was introduced in 1876, adding to the company's existing line, which included pickles, horseradish, and sauerkraut.

184 **"I shall be glad":** AGB to Mabel Hubbard, June 21, 1876.

185 **First, Bell missed the application deadline:** Bruce, *Bell,* p. 190. See also MacKenzie, *Alexander Graham Bell,* p. 119.

185 **As Bell's daughter Elsie recounted:** Elsie Grosvenor, "Mrs. Alexander Graham Bell—A Reminiscence," *Volta Review,* vol. 59 (1957), pp. 209–305. See also Gray, *Reluctant Genius,* p. 134.

185 **Bell explains in a letter to his mother:** AGB to Eliza Bell, June 18, 1876.

185 **"My darling May":** AGB to Mabel Hubbard, June 21, 1876.

186 **"It was very hard":** Mabel Hubbard to AGB, June 19, 1876.

186 **an inconvenient time:** See, e.g., Waite, *Make a Joyful Sound,* p. 131.

186 **"Mr. Hubbard assured him":** MacKenzie, *Alexander Graham Bell,* p. 119.

187 **"I must say I don't like this":** AGB to Mabel Hubbard, June 21, 1876.

187 **"very hopeless":** Ibid.

188 **The judges had chosen:** AGB to his parents, June 22, 1876.

188 **swelteringly hot:** See Hounshell, "Bell and Gray," *Proceedings of the IEEE,* p. 1305.

188 **rotund Dom Pedro II:** Grosvenor and Wesson, *Alexander Graham Bell,* p. 72.

188 **gave an impressive demonstration:** Hounshell, though Bell was dismissive. See AGB to his parents, June 27, 1876.

188 **Professor George F. Barker:** AGB to his parents, June 27, 1876.

188 **"Home, Sweet Home":** Grosvenor and Wesson, *Alexander Graham Bell,* p. 72.

189 **the emperor now greeted him warmly:** AGB to his parents, June 27, 1876. See also Bruce, *Bell,* p. 194.

189 **the remote East Gallery:** Bruce, *Bell,* p. 195.

189 **"I then explained":** AGB to his parents, June 27, 1876.

190 **a membrane transmitter:** Bruce, *Bell,* pp. 196–98.

191 **"Where is Mr. Bell?":** As recounted in AGB to his parents, June 27, 1876.

191 **"I hear, I hear!":** Ibid.

191 **"The Emperor had just":** Gray testimony in *Dowd* 108, p. 138.

191 **"At the Centennial Mr. Bell exhibited":** Quoted in Horace Coon, *American Tel & Tel: The Story of a Great Monopoly* (New York: Longmans, Green, 1939), p. 54.

192 **"a scientific toy":** Elisha Gray to W. D. Baldwin, November 1, 1876, Elisha Gray Collection, Archive Center, National Museum of American History, cited in Hounshell, "Bell and Gray," *Proceedings of the IEEE,* p. 1312.

192 **"Bell has talked so *much*":** Elisha Gray to A. L. Hayes, August 15, 1876, Elisha Gray Collection, Archive Center, National Museum of American History, cited in Hounshell, "Elisha Gray and the Telephone," *Technology and Culture,* p. 157.

193 **explained to the *New York Times*:** See ibid., p. 145. *New York Times,* July 10, 1874, quoted Western Union official Albert Brown Chandler.

193 **"Recently, and long since":** See letter from Elisha Gray, published in *Electrical World and Engineer,* February 2, 1901, p. 199.

193 **"A list of the scientists":** Thomas A. Watson, "The Birth and Babyhood of the Telephone: An Address Delivered Before the Third Annual Convention of the Telephone Pioneers of America, Chicago, October 17, 1913," p. 22.

194 **The *Boston Advertiser*:** "Telephony: Audible Speech Conveyed Two Miles by Telegraph," *Boston Advertiser,* October 9, 1876. The article lists the "Boston Record" of the conversation, as transcribed by Watson, beside the "Cambridgeport Record" as transcribed by Bell. An excerpt is reprinted in Boettinger, *The Telephone Book,* p. 89.

195 **quietly approached Western Union:** According to Watson, "Birth and Babyhood," p. 23. See also W. Bernard Carlson and Michael E. Gorman "A Cognitive Framework to Understand Technological Creativity: Bell, Edison, and the Telephone," in Robert Weber and David Perkins, eds., *Inventive Minds: Creativity in Technology* (New York: Oxford University Press, 1992), p. 64.

195 **Lyceum Hall:** "The Telephone: More Interesting Experiments Between Boston and Salem," *New York Times* (from the *Boston Transcript*, Feb. 24), February 27, 1877, p. 5.

195 **Gertrude Hubbard protested:** As recounted by Mabel Hubbard in a letter to AGB, April 6, 1877.

195 **that evening's proceeds:** Speech by AGB, November 2, 1911.

195 **a silver brooch:** A photograph of the brooch can be seen in Grosvenor and Wesson, *Alexander Graham Bell,* p. 78.

195 **It was a modest wedding:** Bruce, *Bell,* p. 233.

197 **As a wedding present:** Ibid.

16: Conference Call

199 **used to explore the ocean floor:** Claire Calcagno, "Edgerton's Work on Underwater Archaeology," Dibner Institute Colloquium, MIT, November 2, 2004.

199 **My title:** Seth Shulman, "Did Bell Steal the Telephone?," Dibner Institute Colloquium, MIT, February 15, 2004.

201 **an extended fifteen-month trip:** The newlyweds sailed from New York on the *Anchoria* on August 4, 1877, and returned to Quebec, Ontario, on November 10, 1878—Mabel Hubbard Bell to Mrs. Alexander Melville Bell, August 4, 1877; see also Bruce, *Bell,* p. 235.

201 **more dignified without the "k":** Grosvenor and Wesson, *Alexander Graham Bell,* p. 66.

201 **Rhode Island–based entrepreneur:** Bruce, *Bell,* p. 231.

202 **"Of one thing I am quite determined":** AGB to Mabel Hubbard Bell, September 9, 1878.

202 **claiming the legal cover:** Bruce, *Bell,* pp. 262–63.

203 **The legal rules of the day:** Gardiner Hubbard to AGB, November 2, 1878.

203 **threatened with having to forfeit:** Watson, *Exploring Life,* pp. 151–52.

203 **wrote Bell's father for help:** Gardiner Hubbard to Alexander Melville Bell, November 2, 1878.

203 **Hubbard even dispatched Watson:** Watson, *Exploring Life,* p. 152.

203 **"I found Bell even more dissatisfied":** Ibid.

203 **"run the risk of losing him":** Ibid.

204 **"Oh, if I could only":** Mabel Hubbard to Gertrude Hubbard, November 1878, quoted in Toward, *Mabel Bell,* p. 59.

204 **reprise his role in court:** Bruce, *Bell,* p. 270.

204 **to Queen Victoria:** See "The Telephone at Court," *The Times* (London), January 16, 1878.

204 **50,000-franc Volta Prize:** See "The Volta Prize of the French Academy Awarded to Prof. Alexander Graham Bell," *Daily Evening Traveler* (Boston), September 1, 1880. The article concludes that, on the basis of the prize, "The rival claims as to priority of invention may now be regarded as disposed of once and for all. . . ."

205 **a lavish retreat:** For a history of the Bell's estate, see Mabel Hubbard Bell, "The Beinn Bhreagh Estate," typewritten history in *Beinn Bhreagh Recorder,* February 14, 1914, pp. 125–38.

205 **"his life has been shipwrecked":** Quoted in Wesson and Grosvenor, *Alexander Graham Bell,* p. 113.

205 **"all through [Bell's] life":** American Telephone and Telegraph Co., *Alexander*

Graham Bell: Inventor of the Telephone (New York: American Telephone & Telegraph Co., 1947), p. 6.

206 **"Why should it matter":** AGB to Mabel Hubbard Bell, August 21, 1878.

207 **"I became convinced":** Letter from Elisha Gray, published in *Electrical World and Engineer,* February 2, 1901, p. 199.

207 **"Gray, you invented":** Ibid.

207 **"The history of the telephone":** Note found by Lloyd W. Taylor, in Lloyd W. and Ester B. Taylor Papers, Oberlin College Archives.

208 **rules of discovery:** In 1938, civil procedure, including the legal rules of discovery, was reformed with the adoption of the Federal Rules of Civil Procedure. For more, see Charles Alan Wright, *Federal Practice and Procedure* (St. Paul, MN: West Publishing Co., 1969).

209 **the Bell family papers:** Author's interview with Leonard Bruno, curator at the U.S. Library of Congress, November 2006.

210 **"Though independently attested records":** Bruce, Foreword, Grosvenor and Wesson, *Alexander Graham Bell,* p. 6.

211 **Emile Berliner and Thomas Edison:** Emile Berliner, "Improvement in Telephones," U.S. Patent 199,141, issued January 15, 1878, and Thomas A. Edison, "Improvement in Speaking-Telegraphs," U.S. Patent 203,015, issued April 30, 1878. For a discussion, see John Brooks, *Telephone: The First Hundred Years* (New York: Harper & Row, 1975), pp. 70–71.

211 **built flying machines:** See Bell, *Beinn Bhreagh Recorder,* an ongoing record Bell kept of his work while at his estate in Canada. Aviation and sheep breeding are both mentioned, e.g., in the volume of the *Recorder* dated July 24, 1909, to October 19, 1909.

211 **the Mohawk tribe:** AGB, Laboratory Notebook, "From undated to April 23, 1903," LOC (Subject File Folder: The Deaf, Visible Speech, Mohawk Language, 1870–1903).

212 **"a sort of greenhouse effect":** Quoted in Grosvenor and Wesson, *Alexander Graham Bell,* p. 274.

CREDITS

Unless otherwise listed, all illustrations are from the Library of Congress, Prints and Photographs Division, Gilbert H. Grosvenor Collection of Photographs of the Alexander Graham Bell Family. Library of Congress (LC) negative serial numbers are listed.

Page 12: From *The Bell Telephone: The Deposition of Alexander Graham Bell in the Suit Brought by the United States to Annul the Bell Patents* (Boston: American Bell Telephone Co., 1908) (cited hereafter as *Deposition of Alexander Graham Bell*).
Page 23: Library of Congress (LC), Alexander Graham Bell Family Papers (cited hereafter as AGBFP), Laboratory Notebook, 1875–1876.
Page 27: LC-G9-Z1-14931-A.
Page 28: LC-USZ62-53877.
Page 33: Courtesy of the Smithsonian Institution.
Page 36: LC, AGBFP, Laboratory Notebook, 1875–1876.
Page 37: (top) Gray Caveat, U.S. Patent and Trademark Office (cited hereafter as USPTO).
Page 37: (bottom) Amalgam (made by the author).
Page 40: LC-G9-Z1-131,487-A.
Page 41: LC-G9-Z1-131489-A.
Page 47: *Deposition of Alexander Graham Bell.*
Page 53: LC-G9-Z2-4429-B-3.
Page 55: Courtesy of AT&T Photo Archive.
Page 57: LC-USZ62-112820.

Page 59: Courtesy of the photo archive at the Alexander Graham Bell National Historic Site (cited hereafter as AGBNHS), Baddeck, Nova Scotia.
Page 61: Courtesy of AT&T Photo Archive.
Page 66: LC-USZ62-105888.
Page 81: From Daniel Davis, Jr., *Davis' Manual of Magnetism* (Boston: Daniel Davis, Jr., 1847).
Page 86: LC-G9-Z1-14358-A.
Page 101: USPTO.
Page 120: *Deposition of Alexander Graham Bell.*
Page 131: Courtesy of AT&T Photo Archive.
Page 143: Courtesy of Oberlin College Archive.
Page 152: LC, AGBFP.
Page 180: Courtesy of Philadelphia Free Library.
Page 181: LC-USZ62-57385.
Page 183: LC-USZ62-96109.
Page 190: LC-G9-Z4-68813-T.
Page 196: AGBNHS Photo Archive.
Page 197: (left) LC-G9-Z1-144,963-A; (right) LC-G9-Z1-156,508-A.
Page 206: LC-G9-149066-A.

INDEX

Page number in *italics* refer to illustrations.